Teacher Edition

Writer's Companion

Support and Practice for Writing

Grade 3

Printed in the United States of America

ISBN 0-15-349534-0

4 5 6 7 8 9 10 073 15 14 13 12 11 10 09 08 07

Contents

Introduction 4

UNIT 1
WRITER'S CRAFT: VOICE

LESSON 1 Sensory Details 8
Literature Model Officer Buckle and Gloria
Student Model Descriptive Paragraph
LESSON 2 Expressing Your Viewpoint 14
Literature Model Pepita Talks Twice
Student Model Journal/Diary Entry
LESSON 3 Expressing Feelings 20
Literature Model Nate the Great: San Francisco Detective
Student Model Personal Narrative
Test Prep **LESSON 4 Review/Extended Written Response** 26
Literature Model Allie's Basketball Dream
Student Model Story
Test Prep **LESSON 5 Writing Test Practice** 34

UNIT 2
WRITER'S CRAFT: ORGANIZATION

LESSON 1 Time-Order Words and Sequencing 38
Literature Model Turtle Bay
Student Model Directions
LESSON 2 Topic Sentence and Details 44
Literature Model Balto, the Dog Who Saved Nome
Student Model Paragraph of Information
LESSON 3 More About Topic Sentence and Details 50
Literature Model Wild Shots, They're My Life
Student Model Summary
Test Prep **LESSON 4 Review/Extended Written Response** 56
Literature Model Sue, the Tyrannosaurus Rex
Student Model How-to Essay
Test Prep **LESSON 5 Writing Test Practice** 64

UNIT 3
WRITER'S CRAFT: IDEAS

LESSON 1 Supporting an Opinion 68
Literature Model The Stories Julian Tells
Student Model Persuasive Paragraph
LESSON 2 Staying on Topic 74
Literature Model The Talent Show
Student Model Speech
LESSON 3 Reasons and Examples 80
Literature Model Centerfield Ballhawk
Student Model Persuasive Letter
Test Prep **LESSON 4 Review/Extended Written Response** 86
Literature Model Ramona Forever
Student Model Review
Test Prep **LESSON 5 Writing Test Practice** 94

UNIT 4
WRITER'S CRAFT:
SENTENCE FLUENCY

LESSON 1 Combining Sentences **98**
Literature Model Papa Tells Chita a Story
Student Model Paragraph that Compares
LESSON 2 Sentence Variety **104**
Literature Model Coyote Places the Stars
Student Model Paragraph that Contrasts
LESSON 3 Sentence Variety **110**
Literature Model Why Mosquitos Buzz in People's Ears
Student Model Paragraph that Explains
Test Prep **LESSON 4 Review/Extended Written Response** **116**
Literature Model Lon Po Po
Student Model Compare and Contrast Essay
Test Prep **LESSON 5 Writing Test Practice** **124**

UNIT 5
WRITER'S CRAFT:
ORGANIZATION

LESSON 1 Topic and Notes **128**
Literature Model Leah's Pony
Student Model Notes
LESSON 2 Organizing Information **134**
Literature Model Yippee-Yay!
Student Model Outline
LESSON 3 Writing a Research Draft **140**
Literature Model Boom Town
Student Model Research Draft
Test Prep **LESSON 4 Review/Extended Written Response** **146**
Literature Model Cocoa Ice
Student Model Edited Research Report
Test Prep **LESSON 5 Writing Test Practice** **154**

UNIT 6
WRITER'S CRAFT:
WORD CHOICE

LESSON 1 Imagery and Figurative Language **158**
Literature Model I'm in Charge of Celebrations
Student Model Poem
LESSON 2 Vivid Verbs **164**
Literature Model Alejandro's Gift
Student Model Thank-You Letter
LESSON 3 Vivid Verbs and Specific Nouns **170**
Literature Model Rocking and Rolling
Student Model Play
Test Prep **LESSON 4 Review/Extended Written Response** **176**
Literature Model The Armadillo from Amarillo
Student Model Invitation
Test Prep **LESSON 5 Writing Test Practice** **184**

CONVENTIONS

Writer's Grammar Guide **188**
Proofreading Strategies **190**
Proofreading Checklist/Proofreader's Marks **190**

WRITER'S RESOURCES

Presenting Your Work **191**
Writer's Glossary of Terms **191**
Rubrics **192**
Graphic Organizers **193**
Assessment **199**

Introduction

The *Writer's Companion* is designed to deepen students' understanding of the elements and traits of effective writing and to explicitly show students *how* to incorporate those elements and traits into their own writing. It is also designed to deliver carefully targeted instruction in the more difficult tasks that are associated with the writing process—in particular, *organizing, evaluating,* and *revising.*

The *Writer's Companion* includes the following features:

- short literature excerpts that serve as writing models for writer's craft skills
- student writing models as vehicles for teaching the crafts of evaluating and revising
- explicit instruction in writer's craft and the traits of good writing
- some grammar practice (sentence construction and common usage errors)
- practice using rubrics to evaluate writing
- practice with multiple-choice formats encountered in state writing assessment tests

There are 30 *Writer's Companion* lessons per grade. The lessons have different lengths and focuses, as follows:

Lesson	Length	Focus
Lesson 1*	6 pages	Pages 1–3: Writer's craft focus, including a Literature Model Page 4: Example of a writing form (Student Model) Pages 5–6: Evaluating and revising (incorporates both the craft and the form)
Lesson 2	6 pages	Same as for Lesson 1
Lesson 3	6 pages	Same as for Lessons 1 and 2
Lesson 4	8 pages	Pages 1–3: Review of writer's craft skills from Lessons 1–3 Page 4: Example of a writing form (Student Model) Pages 5–6: Evaluating two Student Models using a rubric (incorporates both the crafts and the form) Pages 7–8: Extended writing/Test Prep (self-selected topic)
Lesson 5	4 pages	Pages 1–4: Test Prep (multiple-choice questions)

*Lesson 1 refers to the first lesson in a unit, Lesson 2 refers to the second lesson in the unit, and so on.

A Closer Look at the Lessons

LESSONS 1–3 *(6 pages each)*

Pages 1–3 of each lesson use a Literature Model to teach a writer's craft skill.

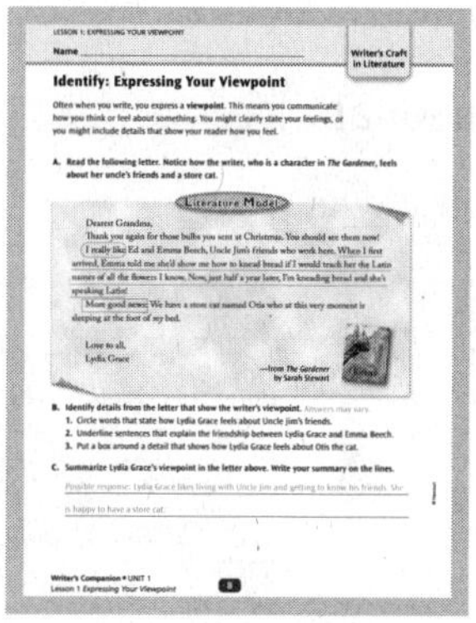
Identify: Expressing Your Viewpoint

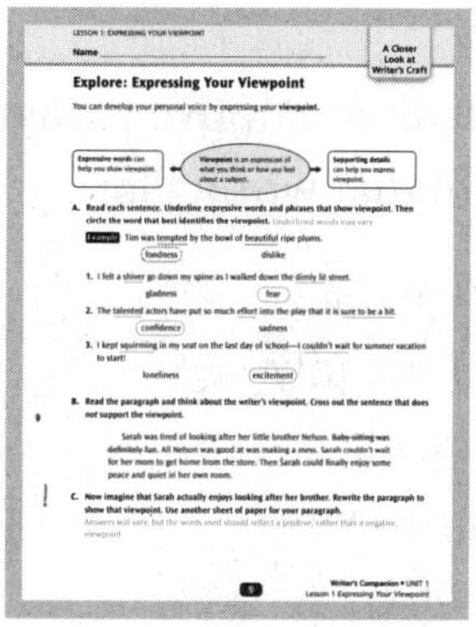
Explore: Expressing Your Viewpoint

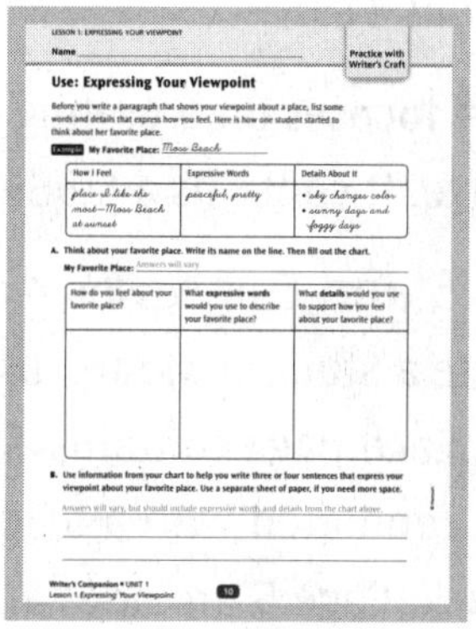
Use: Expressing Your Viewpoint

Page 4 of each lesson teaches a writing form. It contains a Student Model that represents the form (personal narrative, character sketch, etc.) and incorporates the writer's craft skill. Students are prompted to identify examples of the writer's craft in the model, and to identify some of the organizational elements particular to the writing form.

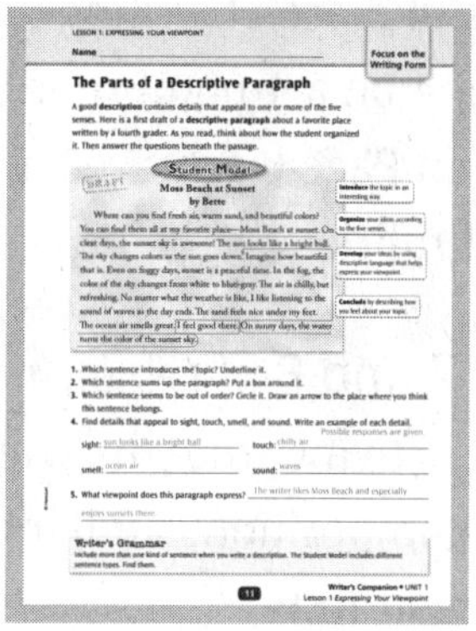
The Parts of a Descriptive Paragraph

Page 5 of each lesson focuses on evaluating, using the Student Model on the previous page.

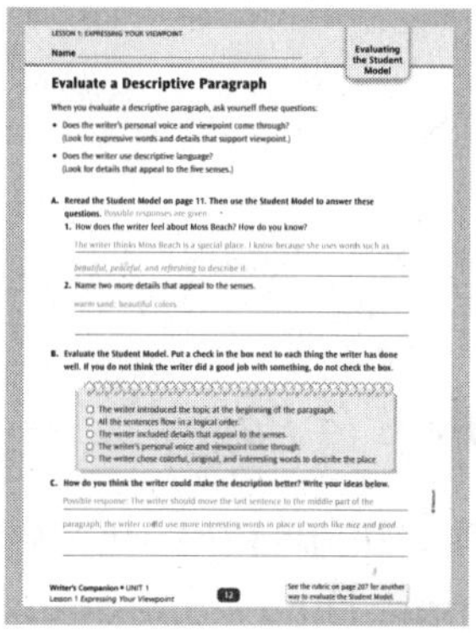
Evaluate a Descriptive Paragraph

Page 6 of each lesson focuses on revising. Students are guided to revise portions of the Student Model, with a focus on writer's craft.

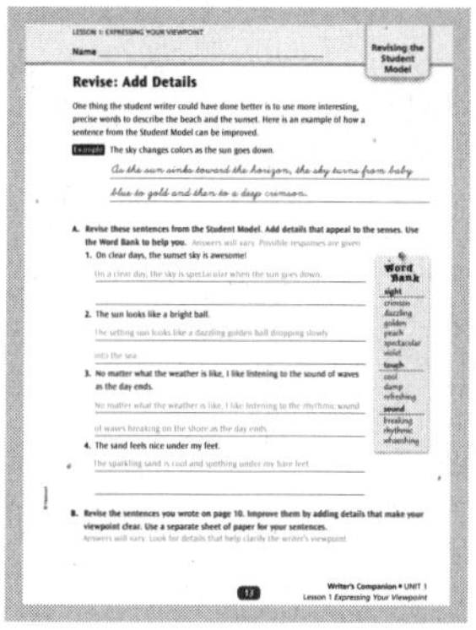
Revise: Add Details

A Closer Look at the Lessons

LESSON 4 *(8 Pages)*

Pages 1–3 of Lesson 4 review all of the writer's craft skills that were taught in Lessons 1–3.

Page 4 focuses on a writing form, such as descriptive paragraph or personal narrative. It contains a Student Model of that writing form.

Pages 5 and 6 of Lesson 4 allow students to evaluate a Student Model, using a child-friendly rubric. Each page contains a longer Student Model annotated with a teacher's constructive feedback. Page 5 provides an example of a successful piece of writing that achieved a score of 4 on a 4-point rubric. The Student Model on Page 6 achieved a score of 2 on a 4-point rubric.

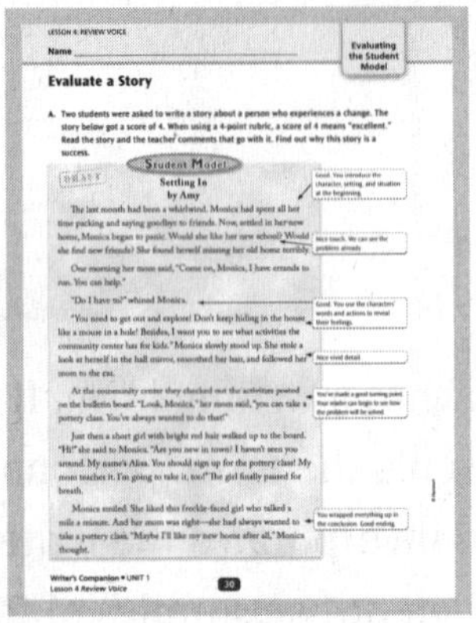

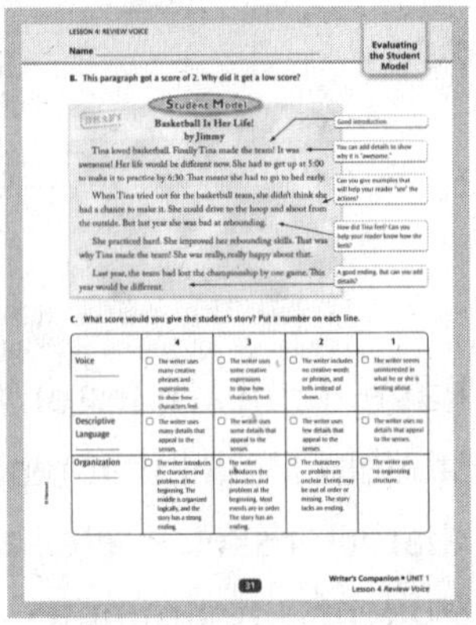

Pages 7 and 8 of Lesson 4 contain an open-ended, extended writing activity. Students select the topic and complete a full-length composition, similar to tasks encountered on state writing assessment tests.

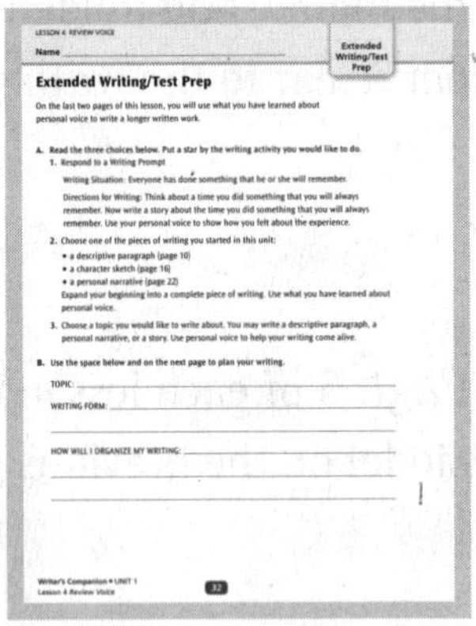

LESSON 5 *(4 Pages)*

Lesson 5 helps prepare students for taking tests. It contains 4 pages of multiple-choice questions that focus on usage, capitalization, and other skills typically encountered in state writing assessment tests.

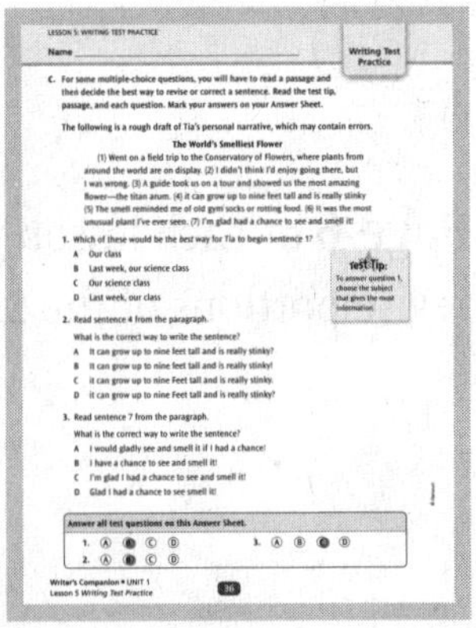

Introduction

When you first learn a new game, such as tennis or baseball, you usually are not very good at it. The more you play, the better you can become.

You can also get better at writing by doing it. This book will give you the skills, strategies, tips, and models you need to become the best writer you can be.

The Writing Process

Writing is a process in which you try different things and go through different steps. The writing process is often divided into five stages. Most writers go back and forth through the stages.

Prewriting

In this stage, you plan what you're going to write. You choose a topic and brainstorm ideas about it. You think of a good order for the ideas.

Drafting

In this stage, you put your ideas in writing as sentences and paragraphs. Follow your Prewriting plan to write a first draft.

Revising

In this stage, you may work by yourself or with a partner or group. Look over your writing, and see how you can make it clearer and stronger.

Proofreading

In this stage, you polish your work. Check for mistakes in grammar, spelling, capitalization, and punctuation. Make a final copy of your composition.

Publishing

Finally, you choose a way to present your work to others. You may want to add pictures, make a class book, or read your work aloud.

Writer's Craft and Writing Traits

You know that to play a game well, you need to use special skills and strategies. In baseball, for example, a player needs to hit well, catch well, and run quickly.

Good writing takes special skills and strategies, too. This web shows the traits, or characteristics, of good writing. You'll learn much more about these traits in this book.

The Traits of Good Writing

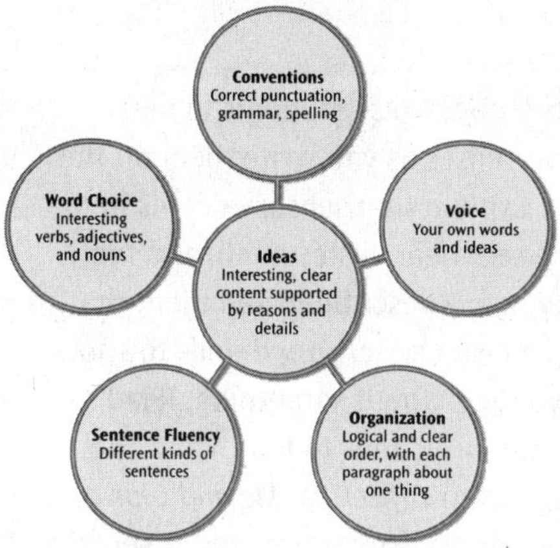

Traits Checklist

As you practice writing, ask yourself these questions.

☑ FOCUS/IDEAS	Is my purpose clear? Do I stay on the topic? Do I use details to support my ideas?
☑ ORGANIZATION	Are my ideas in a clear order? Do I have a beginning, a middle, and an ending? Are my ideas grouped in paragraphs? Do I use transitions, such as time-order words?
☑ VOICE	Do I use my own words and ideas? Do I seem to care about my topic and my audience?
☑ WORD CHOICE	Do I use specific nouns, strong verbs, and colorful adjectives?
☑ SENTENCE FLUENCY	Do I use different kinds of sentences?
☑ CONVENTIONS	Are my spelling, grammar, and punctuation correct?

Look at Sensory Details

OBJECTIVES

- To understand the terms *sensory details* and *viewpoint*
- To identify sensory details and viewpoint in literature

Teach/Model Ask students to name the five senses as you write them on the board. Explain that sensory details help people see, hear, taste, smell, and feel what you are describing. Tell them that you can also use sensory details to show how you feel about something. Read aloud the directions to Part A and the passage from *Officer Buckle and Gloria*. Have students listen for words or sensory details that help them picture the children in the audience.

Guided Practice Help students identify a few sensory details the writer uses to help them picture the children in the audience. Then, after reading aloud the directions for part B, guide students to complete it.

Independent Writing Practice Remind students that sensory details can be used to show feelings. After reading the directions aloud for Part C, have students complete it independently, using examples from the passage.

Use a Literature Model

LESSON 1: SENSORY DETAILS

Name ______________________

Writer's Craft in Literature

Look at Sensory Details

Sensory details help the reader hear, see, smell, taste, or feel what a writer is describing. Writers also use sensory details to show their feelings or **viewpoint.**

A. Read this passage. Notice how the writer uses sensory details to describe a big crowd.

Literature Model

Officer Buckle thought of a safety tip he had discovered that morning.

"NEVER leave a THUMBTACK where you might SIT on it!"

The audience roared. Officer Buckle grinned. He said the rest of the tips with *plenty* of expression.

The children clapped their hands and cheered. Some of them laughed until they cried."

—from *Officer Buckle and Gloria*
by Peggy Rathman

B. Find the sensory details that make the story come alive. Some words will appeal to more than one sense. Possible answers are given.

1. Circle the words that describe what you can hear.
2. Put a box around words that describe what you can see.
3. Underline words that describe what you can feel or touch.

C. What words in the passage tell you that the audience enjoyed watching Officer Buckle and Gloria? Write them below.

Possible response: The audience roared. The children cheered and clapped. Some laughed until they cried.

Writer's Companion ▪ UNIT 1
Lesson 1 *Sensory Details*

8

SHARING AND DISCUSSING

Have students discuss how they were able to tell that the audience enjoyed watching Officer Buckle.

Invite students to read their answers to Part C aloud.

Explore Sensory Details

OBJECTIVES

- To deepen students' understanding of sensory details and the senses to which they appeal
- To explore using sensory details and the senses to which they appeal

Teach/Model Use the concept web at the top of the page to help students understand why writers use sensory details.

Guided Practice Tell students that by using sensory details effectively, writers can show how they feel about a particular subject. Writers can also use sensory details to show how the characters in their stories feel. Read the directions for Part A aloud. Then model how to complete the example, and work through the first sentence with students.

Independent Writing Practice Have students complete the rest of Part A independently. Then, after reading aloud directions for Part B, have them go on to it. For instructions on Part C, see Extending the Concept below.

LESSON 1: SENSORY DETAILS

Name ____________________

A Closer Look at Writer's Craft

Explore Sensory Details

Use **sensory details** to paint a clear picture when you write. Sensory details also help the reader know how you feel about your topic.

Sensory details help the reader see, hear, taste, smell, and feel what you are describing. → **Sensory details** can help you develop a strong personal voice in your writing. ← **Sensory details** help the reader know how you feel about your topic.

A. Read each sentence. Then circle the senses that the details appeal to.

Example I gobbled up the sweet, juicy cherries.

sound (taste) sight

1. I frowned at the gray, rainy sky.

   (sight) touch smell

2. The sweet scent of apple pie filled the house.

   (smell) sight touch

3. My head sunk into the soft pillow, and I fell fast asleep.

   (touch) smell sight

B. Read this sentence and underline the sensory details. Then answer the question.

The slow rocking of the boat made Shawna sit back and relax.

How does Shawna feel? calm, peaceful

C. Now imagine that Shawna gets sick on boats. Rewrite the sentence to show that she does not like being on the boat.

Possible response: The endless rocking of the tiny boat made Shawna groan.

9

Writer's Companion • UNIT 1
Lesson 1 *Sensory Details*

EXTENDING THE CONCEPT: REVERSING VIEWPOINT

Before having students complete Part C, discuss their responses to Part B. Ask volunteers to share which sensory details they underlined and to describe how Shawna feels on the boat. Then ask students how they could change the sentence to show that the boat made Shawna feel ill. Help students identify specific words in the sentence that would need to change. Then, after reading aloud directions for Part C, have students complete it independently. When they are finished, ask students to read aloud their new sentences and compare the details with those in the old sentence.

Use Sensory Details

OBJECTIVES

- To prepare to write by selecting a topic and generating sensory details
- To practice using sensory details

Teach/Model Tell students that before they write a description, they should spend some time planning what they want to say and jotting down some details that they can include. Have students track the text as you read aloud the introduction at the top of the page. Then have volunteers read the details in the example.

Guided Practice Read aloud the directions for Part A. Then have students write the name of the sport they chose on the line. Guide students to use expressive and descriptive words (sensory details) as they complete their charts.

Independent Writing Practice After reading aloud the directions for Part B, have students complete Part B independently. Invite them to read their sentences aloud.

LESSON 1: SENSORY DETAILS

Name ______________________

Practice with Writer's Craft

Use Sensory Details

Before you write about a sport, list some details that tell what you can see, taste, hear, smell, and feel at the game. Here is how one student started to think about her soccer game.

Example

See	Taste	Hear	Smell	Touch
• many people in the stands	• buttery	• loud cheering • whistle from referee	• popcorn	• clapped • retied

A. **Think about a sport you have watched or played. Write its name on the line. Then fill out the chart.**

Name of Sport: Answers will vary.

See	Taste	Hear	Smell	Touch

B. **Use information from your chart to write three sentences about a sport you have watched. Do your writing on another sheet of paper.** Answers will vary, but should include expressive words and details from the chart above.

Writer's Companion • UNIT 1
Lesson 1 *Sensory Details* 10

Reaching All Learners

BELOW LEVEL

Have students who are writing below grade level work with a partner or aide and orally describe the sport they chose. The partner or aide can help the student decide what sensory details to write in the chart.

ADVANCED

Encourage students writing above grade level to use a thesaurus to help them choose more specific, expressive words to describe the sport.

ENGLISH LANGUAGE LEARNERS

Pair less fluent English speakers with more fluent English speakers. Have students discuss the sport they chose and then help each other identify the appropriate English words and phrases to include in the chart.

The Parts of a Descriptive Paragraph

OBJECTIVES

- To understand the parts of a descriptive paragraph
- To analyze a Student Model

Teach/Model Read aloud the introduction and explain that the writing model on this page is a first draft of one student's descriptive paragraph. As students follow along, read aloud the steps for how to organize a descriptive paragraph. Then have students track the text as you read the model aloud. Ask them to pay careful attention to how the details are organized.

Guided Practice Point out the questions below the Student Model. Read the first question aloud and guide students to underline the topic sentence. Explain that a topic sentence gives the main idea of a paragraph. Although a topic sentence is most often found at the beginning of a paragraph, it can also be found in the middle or at the end.

Independent Writing Practice Have students complete Items 2–4 independently. Then discuss where the misplaced sentence belongs.

Use a Student Model

LESSON 1: SENSORY DETAILS

Name ______

Focus on the Writing Form

The Parts of a Descriptive Paragraph

A good **description** contains details that appeal to the senses. Here is a draft of a descriptive paragraph written by a third grader. As you read, think about how the student organized it. Then answer the questions.

Student Model

DRAFT

Goal!
by Ruby

This was the final game of the year. The stands were filled with people. Excited buyers surrounded the popcorn cart. My teammates nervously retied the laces of their cleats and checked their shin guards. They wanted the game to start. The referee threw the ball, and the game began. The crowd cheered loudly and clapped their hands. My friends waved bright, red banners with my jersey number on them. I took a deep breath and crouched down low in front of the goal. I was ready to play my best. I could almost taste the buttery popcorn.

Introduce the topic in the first sentence.

Organize your ideas in an order that makes sense.

Develop your ideas by using sensory details that paint a picture of the scene.

1. Which sentence introduces the topic? Underline it.
2. Which sentence seems to be out of order? Circle it. Then draw an arrow to the place where you think this sentence belongs.
3. Find details that appeal to the senses. Write an example of each detail.

Answers will vary. Possible responses are given.

see: red banners taste: buttery popcorn

hear: cheered touch: retied

4. What do the sensory details tell you about the game?

This was an exciting and important game.

Writer's Grammar
Every sentence should begin with a capital letter and have an end mark. Check to be sure that all sentences in the model are written correctly.

11 Writer's Companion ■ UNIT 1
Lesson 1 *Sensory Details*

WRITER'S GRAMMAR: CAPITAL LETTERS AND PUNCTUATION

Read aloud the Writer's Grammar and ask volunteers to find an example of a sentence within the Student Model that begins with a capital letter and has an end mark. Discuss how the meaning of a sentence might change, depending on the end mark used.

Evaluate a Descriptive Paragraph

OBJECTIVES
- To understand what to look for when evaluating a descriptive paragraph
- To use a checklist to evaluate a descriptive paragraph

Teach/Model Ask students why evaluating (rating) a piece of writing is important. Explain that evaluating helps writers see what works well in their writing and what needs improvement. Read aloud the points at the top of the page and discuss what to look for when evaluating a descriptive paragraph.

Guided Practice Read the directions for Part A aloud. Guide students to look back at the Student Model for information before writing their answers. Explain that a checklist can be a useful way to evaluate writing. The checklist in Part B focuses on many of the writing skills practiced in this lesson.

Independent Writing Practice Read aloud the directions for Parts B and, later, for Part C. Then ask students to complete both parts independently. Explain that they can use their responses to the questions on page 11 to help them.

LESSON 1: SENSORY DETAILS

Name ______________________________

Evaluating the Student Model

Evaluate a Descriptive Paragraph

When you evaluate a descriptive paragraph, ask yourself these questions:

- Does the writer use words that help you picture the scene? (Look for words that appeal to the five senses.)
- Does the writer organize details in a clear way? (Look for an order that makes sense.)

A. Reread the Student Model on page 11. Then follow the directions below.

1. Name two sensory details that help you picture Ruby at the game.

   took a deep breath; crouched down low in front of the goal

2. Give an example of a sensory detail that helps you picture the people in the stands.

   Possible response: friends waved bright red banners

B. Now evaluate the Student Model. Put a check in the box next to each thing the writer did well. If you do not think the writer did a good job, do not check the box.

- ☐ The writer introduced the topic at the beginning of the paragraph.
- ☐ All the sentences flow in a logical order.
- ☐ The writer included details that appeal to the senses.

C. How could the writer make the description better? Write your ideas below.

Possible response: The writer could use more sensory details to describe the start of the game. The last sentence about the popcorn is out of place and needs to be moved.

See the rubric on page 207 for another way to evaluate the Student Model.

SHARING AND DISCUSSING

Have students compare their evaluations of the Student Model. Ask volunteers to share their feelings about being evaluators, what they learned from the activity, and what they would like to do better the next time.

Revise by Adding Details, Examples, and Transition Words

OBJECTIVES

- To revise sentences by adding details and examples
- To revise sentences by including transition words

Teach/Model Tell students that when they revise their writing, they should make changes that improve it. Explain that there are many ways to revise a piece of writing. This time they are going to revise by adding details, examples, and transition words.

Guided Practice Read aloud the introduction and the example revision. Point out that the new sentence has more specific details than the original. After reading aloud the directions for Part A, guide students to use the Word Bank, or their own words, to revise Items 1 and 2 of Part A.

Independent Writing Practice Have students complete the rest of Part A independently. Then, after reading aloud the directions for Part B, have students move on and complete it independently. Invite volunteers to read their revisions aloud.

LESSON 1: SENSORY DETAILS

Name ______________________

Revising the Student Model

Revise by Adding Details, Examples, and Transition Words

One thing the writer could have done better is to add details, examples, and transition words. Transition words such as *soon* or *first* help the reader keep track of ideas. Here is how a sentence from the Student Model can be improved.

Example They wanted the game to start.

One member pulled on her shirt; another fidgeted with her socks while waiting for the game to start.

A. **Revise these sentences from the Student Model. Add details, examples, and transition words. Use the Word Bank to help you.** Answers will vary. Possible responses are given.

1. This was the final game of the year.
   We had reached the championship game, and the winning team would carry home the enormous, shiny trophy.
2. Excited buyers surrounded the popcorn cart.
   Near the loud, crowded stands, excited buyers surrounded the bright, brand-new popcorn cart.
3. The referee threw the ball, and the game began.
   Finally, the referee threw the ball high up into the air, and a player on the green bunted it with her head.

Word Bank

sight
enormous
trophy
shiny
brand-new

touch
reached
carry
bunted

transition words
finally
near

B. **Revise two of the sentences you wrote on page 10. Add details and use interesting, sensory words. If you need more space, use another sheet of paper.**

Answers will vary. Students should add sensory detail words and transitions to make their writing vivid and more interesting.

WRITER'S STRATEGY: REREAD SOFTLY

Tell students that one evaluating strategy they can use is to reread their revision to themselves and make corrections. They should listen for repeated words or phrases and for missing words.

Look at Expressing Your Viewpoint

OBJECTIVES

- To understand the term *viewpoint*
- To identify viewpoint in literature

Teach/Model Explain that writers make a story more lifelike when they include words that tell about feelings and images. Explain that a *viewpoint* is a writer's thoughts and feelings about a subject and that the words a writer chooses can help readers figure out the writer's viewpoint.

Ask students to close their eyes and listen as you read aloud the excerpt from *Pepita Talks Twice*. As you read, encourage students to listen for words and phrases that reflect Pepita's feelings. Tell students that a character's actions and words express their viewpoint.

Guided Practice Reread the passage aloud with the students. Pause to have students role play the actions described in the passage that illustrate Pepita's thoughts and feelings. After reading aloud the directions for Part B guide students in completing it.

Independent Writing Practice After reading aloud the directions for Part C, have students complete it independently.

Use a Literature Model

LESSON 2: EXPRESSING YOUR VIEWPOINT

Name ____________________

Writer's Craft in Literature

Look at Expressing Your Viewpoint

A **viewpoint** is a writer's thoughts and feelings about a subject. One way a writer expresses his or her viewpoint is with colorful words that help the reader know how the writer feels.

A. Read the passage. Notice how Pepita shows her feelings about her decision not to speak Spanish anymore.

Literature Model

That night when she went to bed, Pepita pulled the blankets up to her chin and made a stubborn face. "I'll find a way," she thought. "If I have to I can call myself Pete. I can listen in Spanish. I can hum with the singing. I can call a taco a crispy, crunchy, folded-over, round corn sandwich! And Wolf will have to learn his name!" With that she turned over and went to sleep.

—from *Pepita Talks Twice*
by Ofelia Dumas Lachtman

B. Find the words that the writer uses to show Pepita's viewpoint.

1. Circle the actions showing that Pepita has made up her mind.
2. Underline the colorful words that Pepita uses to describe a taco.

C. How did Pepita feel when she "went to sleep"?

Possible responses: Confident. Happy with what she decided to do. Relieved.

Writer's Companion ■ UNIT 1
Lesson 2 *Expressing Your Viewpoint* 14

EXTENDING THE CONCEPT: A NARRATOR'S VIEWPOINT

Ask who is telling the story in the passage. Ask how the reader knows what Pepita is thinking if she isn't actually speaking. Identify examples of words that show the writer's viewpoint as narrator.

Explore Expressing Your Viewpoint

OBJECTIVES

- To deepen students' understanding of how viewpoint is expressed
- To explore using colorful words to convey emotion

Teach/Model Point out the diagram at the top of the page. Explain that the diagram can help students understand the function of viewpoint and how it relates to expressing thoughts, feelings, opinions, and experiences. Show that it also tells the purpose of using colorful words.

Guided Practice Explain that a writer's words communicate an image, feeling and idea similar to how a photographer snaps a picture. The photographer's viewpoint from the camera becomes the photo. The writer's viewpoint is what the reader reads on a page. Read aloud the directions for Parts A and B. Then guide students in completing them. Discuss how particular words (*oozing, green mold*) show certain feelings (*disgust*.)

Independent Writing Practice For instructions for Part C use Writer's Strategy below.

LESSON 2: EXPRESSING YOUR VIEWPOINT

Name ____________________

A Closer Look at Writer's Craft

Explore Expressing Your Viewpoint

Writers express their **viewpoint** through the words they choose. Colorful words help writers express viewpoints by painting pictures in the reader's mind.

Viewpoint is how a writer feels about his or her subject. → **Express** thoughts, feelings, opinions, and experiences. → Paint pictures in the reader's mind with **colorful words.**

A. **Draw a box around colorful words that "paint a picture" or express a feeling.**

Example I brushed the [soft, feathery snow] off my coat.

1. The rollercoaster ride made [my stomach turn over.]
2. I grew an [oozing, green mold] in science class.
3. I was [safe and warm] under the blanket.

B. **Read this sentence. Then answer the question.**

The beat of the music made the children spring to their feet.

How did the music make everyone feel? Possible responses: like dancing; full of energy

C. **Use colorful words to describe how music makes you feel. You may need to use another sheet of paper.**

Answers will vary, but should describe a feeling.

15 Writer's Companion • UNIT 1 Lesson 2 *Expressing Your Viewpoint*

WRITER'S STRATEGY: FEELING WORDS

Before students complete Part C, have them list different styles of music they might listen to: rock, rap, country, blues, classical, etc. Discuss how each type of music might give them a different feeling. Ask them to consider if liking or disliking a particular song might make a difference in what they would write about how the music makes them feel. Divide the class into small groups. Assign each group a style of music. If possible, provide examples for them to hear. Have the groups brainstorm colorful words for how the music makes them feel. Ask groups to share their responses, as you record them on the board. After reading directions for Part C, explain that students may use these colorful words to help them complete it.

Use Expressing Your Viewpoint

OBJECTIVES

- To prepare to write by using colorful words about a chosen event
- To practice writing with a viewpoint

Teach/Model Tell students that before they write an account of an event, they should ask themselves, "What do I see in my mind when I think about this event? How do I feel?" Read aloud the introduction at the top of the page. Have students look at the example and identify what types of words should go in each column. Explain the distinction between the visual images listed under Colorful Words and the words listed under Thoughts or Feelings.

Guided Practice Read aloud the directions for Part A. Then have students name a recent event in which the class participated (i.e. a class trip, team sport, etc.) and write it on the line provided. Guide them in filling out the chart.

Independent Writing Practice After reading aloud directions for Part B, have students complete it independently.

LESSON 2: EXPRESSING YOUR VIEWPOINT

Name ______________________ Practice with Writer's Craft

Use Expressing Your Viewpoint

In their journals and diaries, writers often express thoughts and feelings about events. They use colorful words. Here is how one student planned his writing about an unusual school day.

Example Event: Show and Tell

Colorful Words	Thoughts or Feelings
black and white pig red leather leash scrambled	scared excited

A. Think about something that happened one day during the past week. Write the topic of the event on the line. Then fill out the chart.

Event: Answers will vary.

Colorful Words	Thoughts or Feelings

B. Use information from your chart to write a journal entry about the event that happened to you. Include colorful words to express your thoughts and feelings. Use another sheet of paper for your writing.

Answers will vary, but should include colorful words and details from the chart.

Reaching All Learners

BELOW LEVEL

Before listing colorful words and feelings in the chart, have students who are writing below grade level to draw a picture of the event using colored pencils. Have them use the images they've chosen to draw to help them think of words to fill the chart.

ADVANCED

Challenge students writing above grade level to change their viewpoint. In addition to writing a journal entry from their personal viewpoint have them write about the event from the viewpoint of a teacher. What information might be revised?

ENGLISH LANGUAGE LEARNERS

Have students who are learning English to tell their stories to you. Then have the students explain how their journal entries should be written as you write them on the board.

The Parts of a Journal/Diary Entry

OBJECTIVES
- To understand the parts of a journal/diary entry
- To analyze a Student Model

Teach/Model Explain to students that the writing model on this page is a student's journal entry about an event written from that student's viewpoint. Read aloud the steps for how to organize a journal entry. Next, read the model aloud. As students listen, have them pay careful attention to the organization of information, the colorful words, and how the thoughts and feelings are expressed.

Guided Practice Point out the questions below the Student Model. Read the first question, or have a volunteer read it, aloud. Guide students to identify and underline the sentence in the model that introduces the topic. Explain that this is called the topic sentence.

Independent Writing Practice Have students complete Items 2 and 3 independently. Then invite them to discuss their responses.

Use a Student Model

LESSON 2: EXPRESSING YOUR VIEWPOINT

Name ______________________

Focus on the Writing Form

The Parts of a Journal/Diary Entry

A **journal** or **diary entry** tells about events that happened. It should use colorful words to help express the writer's viewpoint. Here is a draft of a journal entry written by a third grader. As you read, think about how the student organized the entry. Then answer the questions.

Student Model

DRAFT

Journal Entry
by Darryl

Friday, October 14

An unusual thing happened at school today. As I was finishing my math test, I saw Hans's mother at the door. She held a small black-and-white pig at the end of a red leather leash. I guess she was picking up Hans early. When Hans went to take the leash, he dropped it. The pig ran out of the classroom. The animal must have been scared. We all scrambled to the door to watch. Hans ran after the pig. "Daisy, come back!" he called. It sure would be fun to have a pig come to class more often.

Write the day and date of the entry.

Begin describing the event.

Use colorful words to explain what happened.

Make your viewpoint clear by including your thoughts and feelings about what happened.

1. Which sentence introduces the topic? Underline it.
2. Which sentence sums up Darryl's feelings? Put a box around it.
3. Find colorful words. Write them on the lines below.

Answers will vary. Possible responses: black-and-white, red leather, scrambled to the door

Writer's Grammar
When telling about an event in the past, make sure that all the action words are in the past tense (*was, had*). Look for the action words in the Student Model that show what happened in the past.

17 Writer's Companion ■ UNIT 1
Lesson 2 *Expressing Your Viewpoint*

WRITER'S GRAMMAR: VERB TENSE

Read aloud the Writer's Grammar. Ask volunteers to identify the action words that are written in the past tense. Discuss how using the correct verb tense is important in order to communicate the passage of time. Direct students to think of examples of how the verbs could have been used incorrectly in the model.

Evaluate a Journal/Diary Entry

OBJECTIVES

- To understand what to look for when evaluating a journal/diary entry
- To use a checklist to evaluate a journal entry

Teach/Model Discuss with students why it is important to evaluate writing. Discuss what to look for when evaluating a journal or diary entry and read to students the points at the top of the page.

Guided Practice Read the directions for Part A aloud. Guide students in writing responses and referring to the Student Model on page 17 for information. Point out the checklist in Part B. Tell students that using a checklist is helpful in evaluating writing because it allows them to keep their focus on a few important elements. Then guide them in evaluating whether or not the writer of the Student Model wrote the date on his journal entry.

Independent Writing Practice After reading aloud the directions for Parts B and C, have students complete them independently.

LESSON 2: EXPRESSING YOUR VIEWPOINT

Name ______________________________

Evaluating the Student Model

Evaluate a Journal/Diary Entry

When you evaluate a journal entry, ask yourself these questions:

- Does the writer use colorful words to tell about events, thoughts, and feelings? (Look for the writer's viewpoint.)
- Does the writer organize details in a clear way? (Look for an order that makes sense.)

A. Reread the Student Model on page 17. Then follow the directions below.

1. List examples of colorful words that describe the pig.

   Possible responses: small, black-and-white, scared

2. Give an example of the writer's thoughts or feelings.

   Possible responses: I guess she was picking Hans up early.; The pig must have been scared.

B. Now evaluate the Student Model. Put a check in the box next to each thing the writer has done well. If you do not think the writer did a good job, do not check the box.

- ☐ The day and date appear in the entry.
- ☐ All the sentences flow in a clear order.
- ☐ The writer included colorful words.
- ☐ The writer made his viewpoint clear by expressing thoughts and feelings.

C. How do you think the writer could make the journal entry better? Write your ideas below.

Possible response: The writer could use more colorful words to describe the event.

Writer's Companion ■ UNIT 1
Lesson 2 *Expressing Your Viewpoint* 18

See the rubric on page 207 for another way to evaluate the Student Model.

EXTEND THE CONCEPT: WORD CHOICE

Point out that the viewpoint expressed in the Student Model is very factual, much like a newspaper reporter. Suggest to students that the writer could have used more colorful words to describe how he and his classmates *felt* about the event. Write the following sentences on the board and have students work in pairs to create more "colorful" alternatives for each:

- As I was finishing my math test, I saw Hans's mother at the door.
- Hans ran after the pig.

Revise by Adding Important Details

OBJECTIVES

- To revise sentences by adding important and interesting details

Teach/Model Tell students that revising writing means changing it to make it better. Explain that there are many types of changes writers can make. They can change or revise words, sentences, organization, or ideas. In this activity, students will revise their writing by adding important details.

Guided Practice Read the introduction. Invite one volunteer to read the example and another to read the example revision aloud. Discuss with students how the details improve the sentence. Read aloud the directions for Part A. Then, using the Word Bank and referring to the Student Model on page 17, model how you would revise Item 1 of Part A. Have students use the Word Bank, or their own words, as they revise the remaining items in Part A.

Independent Writing Practice Look at Sharing and Discussing. After reading aloud the directions for Part B, have students complete it independently using their journal entries from page 16.

LESSON 2: EXPRESSING YOUR VIEWPOINT

Name ____________________

Revising the Student Model

Revise by Adding Important Details

One thing the writer could have done better is to add details to make the events clearer and more interesting. Here is one way the first sentence from the Student Model can be improved.

Example An unusual thing happened at school today.

I am usually very eager for Friday afternoon to come, but when the bell rang I didn't even want to move.

A. **Revise these sentences from the Student Model by adding important or interesting details. Use the Word Bank to help you.** Answers will vary. Possible responses are given.

1. I saw Hans's mother at the door.

   Suddenly, Hans's mother was looming in the doorway.

2. When Hans went to take the leash, he dropped it.

   When Hans leaped up to grab hold of the leash, he dropped it.

3. The animal must have been scared.

   I suppose all our shouting terrified the little animal.

Word Bank

grab
leaped
little
looming
suddenly
suppose
terrified

B. **Go ahead and revise the journal entry you wrote on page 16. Add details with colorful words. Do your writing on another sheet of paper.**
Answers will vary. Students should add colorful words to make their writing vivid and more interesting.

19 Writer's Companion ■ UNIT 1
Lesson 2 *Expressing Your Viewpoint*

SHARING AND DISCUSSING

Before students begin Part B, suggest that they use the checklist on page 18 to evaluate their own writing. Once revisions have been made, invite volunteers to read their journal entries aloud. Ask them to share what changes they made when they revised their initial journal entry.

Look at Expressing Feelings

OBJECTIVES

- To understand the term *expressing feelings*
- To identify feelings and how they are communicated in literature

Teach/Model Explain to students that writers make a story more realistic and emotional when they include details that communicate feelings. Explain that writers can express feelings by literally writing what the feeling is. They also can use various words that convey a mood or feeling. They can describe an action that suggests a feeling, as well.

Read the excerpt from *Nate the Great: San Francisco Detective* aloud. Direct students to listen for the words and expressions that communicate feeling.

Guided Practice Reread the passage aloud with the students and read aloud directions for Part B. Then guide students in completing Part B, pausing to have students identify what each character is feeling and how the feelings are expressed in words.

Independent Writing Practice After reading aloud directions for Part C, have students complete it independently.

Use a Literature Model

LESSON 3: EXPRESSING FEELINGS

Name ______________________________

Writer's Craft in Literature

Look at Expressing Feelings

Expressing feelings means telling how you feel about something or someone.

A. **Read the model. Notice how Annie and Nate express their feelings during their phone call.**

Literature Model

"Hello, Nate." It was Annie, from back home.

"We all miss you," she said. "And Fang has something to tell you."

I heard heavy breathing. I knew that Annie's dog, Fang, was on the line. I was happy to be many miles away from his teeth. I waited. Fang had nothing else to say. Then I heard a strange voice. It belonged to Rosamond.

—from *Nate the Great: San Francisco Detective*
by Marjorie Weinman Sharmat & Mitchell Sharmat

B. **Find the words that the writer uses to show how Annie and Nate feel.**

1. Circle the words that show how Annie feels.
2. Underline the words that Nate uses to describe how he feels.

C. **When Annie says, "We all miss you," how do you think she feels?**

Possible responses: Sad; wishing Nate would come home soon

____

Writer's Companion ■ UNIT 1
Lesson 3 *Expressing Feelings*
20

SHARING AND DISCUSSING

Invite students to read the passage aloud as if it were a play, with different students playing the role of Nate and Annie. Discuss how Annie might sound when expressing her feelings to Nate. If Nate was narrating, what tone would he take?

Explore Expressing Feelings

OBJECTIVES

- To deepen students' understanding of how words express feelings
- To explore using "feeling" words to convey an emotion

Teach/Model Use the diagram at the top of the page to help students understand how writers use their personal voice to express their feelings.

Guided Practice Explain to students that a writer's words can directly or indirectly communicate a feeling. Read aloud the directions for Parts A and B. Then guide students in completing them. Discuss with students how particular words can give the reader images (*butterflies in my stomach*) that express feelings (nervous excitement).

Independent Writing Practice After reading aloud directions for Part C, have students complete it independently.

Name ______________________

A Closer Look at Writer's Craft

Explore Expressing Feelings

Writers often express their personal voice by using "feeling" words such as *sad* or *joyful*. Writers also express their personal voice by describing their feelings indirectly.

Feeling

Directly Stated: I feel **sad** that you're away.

Indirectly Stated: I hope you will come home soon. **(misses the person)**

A. Read each sentence. Underline the "feeling" word or words.

Example I was surprised to see so many people at the party.

1. When the thunder woke me up last night, I was scared.
2. I was annoyed because my little brother would just not stop yelling.
3. He was upset since the school bus left without him.

B. Read these sentences. Then write a sentence to describe how the writer feels.

It was the first day at my new school. I had butterflies in my stomach.

Possible response: The writer is nervous about attending a new school.

C. Think of a time you tried something new. Describe what you tried and how you felt about it. Use direct and indirect "feeling" words.

Answers will vary, but should tell about trying something new and use "feeling" words and/or words that describe the feeling.

EXTENDING THE CONCEPT

Use this lesson as a springboard for discussing with students figurative language. Both are excellent examples of expressing feelings in an indirect manner. Challenge students to brainstorm a list of figurative expressions they know. Prompt them to discuss the meaning of these:

- I received a warm reception.
- He was hopping mad.
- She had an icy stare.
- I was walking on air!
- They were blue after they lost the game.

Use Expressing Feelings

OBJECTIVES

- To prepare to write by selecting a topic and writing feelings connected to the topic
- To practice writing feelings connected to the topic

Teach/Model Tell students that before they write an account of an event, they should ask themselves: What happened at this event? How did I feel at the time? How do I feel now? Read aloud the introduction at the top of the page. Direct students to look at the example and identify the columns. Explain the difference between the account of the event listed under "What Happened?" and the words listed under "How Did I Feel?"

Guided Practice Read aloud the directions for Part A. Lead students in filling out the chart.

Independent Writing Practice After reading aloud the directions for Part B, have students complete it independently.

LESSON 3: EXPRESSING FEELINGS

Name ____________________

Practice with Writer's Craft

Use Expressing Feelings

Sometimes a writer tells about an event that really happened in his or her life. The writer also expresses feelings about the event. Here is how one third grader planned to write about a hike in the woods.

Example **Event:** A Hike **People Involved:** David, Justin, and Sam

What happened?	How did I feel?
rode to the park	excited
dropped my lunch in the stream	upset, hungry
bug bites	itchy
wrong turn made the hike longer	tired

A. Think of a trip you have taken. Fill in the event and the people involved. Then complete the chart.

Event: Answers will vary. **People Involved:** Answers will vary.

What happened?	How did I feel?

B. Use the information from your chart to write about your trip. Do your writing on another sheet of paper.
Answers will vary, but should include events and "feeling" words from the chart.

Writer's Companion ▪ UNIT 1
Lesson 3 *Expressing Feelings* 22

Reaching All Learners

BELOW LEVEL

Assist students by structuring the "What Happened?" section. Have students fill in the blanks:
First ...
Next...
Then...
Finally...

ADVANCED

Challenge students writing above grade level to use a range of devices to express their feelings in writing. They may use figurative language, punctuation, etc.

ENGLISH LANGUAGE LEARNERS

Prepare students for this activity by making a "feelings reference chart." Brainstorm a list of words that express a range of emotions. Have students help you write the words in multiple languages.

The Parts of a Personal Narrative

OBJECTIVES

- To understand the parts of a personal narrative
- To analyze a Student Model

Teach/Model Explain to students that the writing model on this page is a first draft of a student's personal narrative. Read aloud the steps for how to organize a personal narrative. Then read the model aloud. As students listen, have them pay careful attention to how the details are organized.

Guided Practice Point out the questions below the Student Model. Read the first question aloud. Then guide students to identify and underline the sentence that introduces the topic. Remind them that this is called the topic sentence.

Independent Writing Practice Have students complete Items 2–3 independently. Then discuss their responses.

Use a Student Model

LESSON 3: EXPRESSING FEELINGS

Name ______________________

Focus on the Writing Form

The Parts of a Personal Narrative

A **personal narrative** tells about an event that happened to the writer. The narrative has a beginning, a middle, and an ending. Here is a draft of a personal narrative written by a third grader. As you read, think about how the student organized the draft. Then answer the questions.

Student Model

DRAFT

The Too-Long Hike
by David

Last Saturday, I went on a hike with Justin and his big brother, Sam. Sam drove us to Woodland Park. I was so excited about going!

On the trail, I tripped and dropped my lunch into the water. I was upset. I didn't have any lunch to eat.

Later, there was a place with lots of bugs. I got so many bug bites. Suddenly, I started itching badly.

We got to a fork in the trail, and Sam led us down the wrong path. It was an hour till we got back to the car. I was tired.

That was one hike I wasn't sorry to see end. It had been a too-long hike.

Introduce the topic. Write a topic sentence telling what your personal narrative is about.

Tell about the events. Tell only important events that happened on your trip. Write them down in order. Use time-order words to move the narrative along.

Express thoughts and feelings. Include interesting details, sensory descriptions, and words that show what the trip meant to you.

Have a strong ending. Summarize what made your trip special.

1. Which sentence introduces the topic? Underline it.
2. Which sentence tells how the writer felt before the hike? Draw a box around it.
3. Which sentences sum up the narrative? Circle them.

Writer's Grammar
A compound sentence is made up of two smaller sentences. It is joined by a word such as *and, but,* and *or.* Writers use a comma before the joining word. Look for a compound sentence in the Student Model.

23 Writer's Companion ■ UNIT 1
Lesson 3 *Expressing Feelings*

WRITER'S GRAMMAR: COMPOUND SENTENCES

Read aloud the Writer's Grammar. Ask volunteers to identify the compound sentence in the Student Model. Write it on the board:

- We got to a fork in the trail, and Sam led us down the wrong path.

Discuss how the sentence might be revised and written as two sentences. Direct students to identify compound sentences in their own writing. Collect a list of conjunctions such as *and, but,* and *or.* Create a classroom conjunction chart.

Evaluate a Personal Narrative

OBJECTIVES

- To understand what to look for when evaluating a personal narrative
- To use a checklist to evaluate a personal narrative

Teach/Model Remind students the importance of evaluating their writing. Discuss what to look for when evaluating a personal narrative. Use the points at the top of the page.

Guided Practice Read the directions for Part A aloud. Guide students in writing responses and referring to the Student Model on the previous page for information. Point out the checklist in Part B. Explain that students are looking for many of the writing skills practiced in this lesson.

Independent Writing Practice Read aloud the directions for Parts B and C. Then have students complete them independently, referring them back to the Student Model on page 23.

LESSON 3: EXPRESSING FEELINGS

Name ____________________

Evaluating the Student Model

Evaluate a Personal Narrative

When you evaluate a personal narrative, ask yourself these questions:

- Does the writer express his or her feelings?
- Does the narrative have a beginning, a middle, and an ending?

A. Reread the Student Model on page 23. Then answer these questions.

1. Give an example of feelings the writer expresses.

   Possible responses: I was upset. I didn't have any lunch to eat.

2. Why wasn't the writer sorry to see the hike end?

   Possible response: Things went wrong—dropped his lunch, bug bites, and got lost.

B. Now evaluate the Student Model. Put a check in the box next to each thing the writer has done well. If you do not think the writer did a good job, do not check the box.

- ☐ The writer introduced the topic in the first sentence.
- ☐ The writer told about the events in the order that they happened.
- ☐ The writer expressed his feelings clearly.
- ☐ The conclusion summarizes what made the trip special.

C. How do you think the writer could make the personal narrative better? Write your ideas below.

Possible response: The writer could use more details and feeling words to describe the events in the narrative.

Writer's Companion • UNIT 1
Lesson 3 *Expressing Feelings*
24

See the rubric on page 207 for another way to evaluate the Student Model.

EXTEND THE CONCEPT: WORD CHOICE

Suggest to students that the writer could have used more "feeling" words to describe how he *felt* about the hike. Write these sentences on the board. Have students work in pairs to create more words for each:

- Later, there was a place with lots of bugs.
- We got to a fork in the trail, and Sam led us down the wrong path.

Revise by Expressing Feelings Clearly

OBJECTIVES

- To revise by adding "feeling" words
- To revise by adding descriptive details

Teach/Model Explain to students that in this activity they will revise their writing by adding more descriptive words to reflect their feelings.

Guided Practice Read, or have a volunteer read, the introduction and the example aloud. Ask students how the added words improve the sentence by helping to express a feeling. After reading aloud the directions for Part A use the Word Bank to model how you would revise Item 1. Have students use the Word Bank, or their own words, as you revise with them the remaining items in Part A.

Independent Writing Practice Read aloud directions for Part B. Then use Share and Discuss below and have children complete Part B independently.

LESSON 3: EXPRESSING FEELINGS

Name ______________________________

Revising the Student Model

Revise by Expressing Feelings Clearly

One thing the writer could have done better is to add more words that show feelings. Here is an example of how a sentence from the Student Model can be improved.

Example I was so excited about going!

I was so excited about going that I could hardly sit still.

A. **Revise these sentences from the Student Model by adding "feeling" words. Use the Word Bank to help you.** Answers will vary. Possible responses are given.

1. I was upset.

   I was so upset that I punched a tree.

2. I didn't have any lunch to eat.

   I was also so hungry that my stomach growled.

3. Suddenly, I started itching badly.

   Suddenly, I started itching so badly that I wanted to jump out of my skin.

4. It was an hour till we got back to the car.

   The walk back to the car seemed more like a day than an hour.

Word Bank

hungry
jump out of my skin
punched a tree
stomach growled

B. **Revise your writing on page 22. Add words that help you show your feelings clearly. If you need more space, use another sheet of paper.**

Answers will vary. Students should add more "feeling" words to express their feelings more clearly.

SHARE AND DISCUSS

Suggest to students that they use the checklist on page 24 to evaluate their own writing before they make revisions. Once revisions have been made, invite volunteers to read their personal narratives aloud. Ask them to share what "feeling" words they used in their narratives. Write all of the students' "feeling words" on the board. Ask students if there are other words they had considered.

Review Voice

OBJECTIVES

- To review the terms *sensory details, colorful words,* and *words that express feelings*
- To identify how *voice* is expressed using sensory details, colorful words, and words that express feelings and viewpoint

Teach/Model Remind students that for a writer's viewpoint to be expressed clearly, the writer should include sensory details, colorful words, and words that communicate feelings.

Read the excerpt from *Allie's Basketball Dream* aloud. Direct students to listen for the words and details that express the author's viewpoint.

Guided Practice Invite volunteers to name the five senses (*sight, smell, hearing, taste,* and *touch*). Next, after reading directions for Part B aloud, guide students in completing it. Read aloud the model again, pausing to have students identify examples of sensory details as each sentence is read.

Independent Writing Practice Read the directions to Part C aloud. Then have students complete Part C independently. Afterwards, discuss their interpretation of the writer's feelings and what words and details they thought best expressed the writer's viewpoint.

Use a Literature Model

LESSON 4: REVIEW VOICE

Name ______________________

Writer's Craft in Literature

Review Voice

When writing, writers are careful to make their voices clear. They do this by using **sensory details, colorful words,** and **words that express feelings.**

A. Read the following passage. Notice how the writer expresses her voice.

Literature Model

Allie thought it over. She remembered the first time her father took her to a basketball game at Madison Square Garden. She loved it all: The noise of the crowd, the bright lights on the court, and especially the slam-dunks the players made look so easy! She knew right then and there that one day, she would be a professional basketball player too…

—from *Allie's Basketball Dream* by Barbara E. Barber

Possible answers are given.

B. Reread the passage. Look for sensory details that express the writer's voice clearly.

1. Underline a detail that tells about sight.
2. Circle a detail that tells about hearing.
3. Put a box around a detail that tells about touch.

C. How does the writer feel about basketball? Use details from the passage to help you answer.

Possible response: The writer loves basketball. She says Allie "loved it all," and her description sounds as if she does, too.

Writer's Companion • UNIT 1
Lesson 4 *Review Voice*

26

WRITER'S GRAMMAR: ADJECTIVES

Tell students that it is often a writer's *choice* of *adjectives* that expresses his or her viewpoint most clearly (*blue, happy, oldest*). Explain that an adjective usually comes before a noun or pronoun that it describes.Guide students in identifying and underlining the adjectives in the model.

Review Voice

OBJECTIVES

- To review how writers use voice to show what characters are like and how they feel
- To practice identifying sensory details, colorful words, and "feeling" words

Teach/Model Read aloud the introduction. Then use the graphic organizer at the top of the page to review the different ways a writer can show how characters feel and what characters are like.

Guided Practice Read aloud the directions for Part A. Next, read the passage aloud. Ask students to describe Brianna. Reinforce that the writer uses details, colorful words, and "feeling" words to show what the character is like and how she feels. Read aloud the directions for Part B and the example. Then guide students in completing Item 1.

Independent Writing Practice Have students complete the rest of Part B. Then read aloud directions for Part C and have students complete it independently.

LESSON 4: REVIEW VOICE

Name ____________________

A Closer Look at Writer's Craft

Review Voice

Writers use **voice** to show what characters are like and how they feel.

VOICE

- **Sensory details** appeal to the five senses.
- **Colorful words** "paint a picture" to express a viewpoint.
- **"Feeling"** words tell how writers and characters feel.

A. Read this beginning part of a story. Look for words and details that show what the characters are like and how they feel.

Brianna sniffed the air as the muffins baked. They smelled so good her mouth watered. Only ten more minutes to wait. She could almost taste the sweet, tangy blueberries. She put on a CD to make the time go faster. The beat was loud and thumping. Brianna danced joyfully around the room. The music was so loud that she didn't hear the timer go off. She danced and danced, lost in the music. Suddenly, she stopped. A strong, smoky smell made her cough. Black smoke was coming from the kitchen. Something was on fire!

B. Write the details that appeal to each sense.

Example **smell:** They smelled so good her mouth watered.

1. **hearing:** Possible response: The beat was loud and thumping.
2. **smell:** Possible response: A strong, smoky smell made her cough.
3. **sight:** Possible response: Black smoke was coming from the kitchen.

C. How does Brianna feel when she is dancing? Which colorful and "feeling" words tell you this?

Possible response: She feels happy. You this know because the writer uses the feeling word "joyfully" and the colorful words "lost in the music."

EXTENDING THE CONCEPT: WORD CHOICE

Explain to students that the words a writer chooses, even when not specifically describing the character, can still reflect what a character is like or how they feel. Point out how the descriptions of the food (*sweet, tangy blueberries*) and music (*loud and thumping*) in the passage are similar to Brianna because she is in a good mood. Suggest to students that if Brianna was in a sad mood that the descriptions of the baking muffins, the sound of the music, and how Brianna danced might have been different. As a class or in small groups, work with students to rewrite the paragraph as if Brianna were sad.

Review Voice

OBJECTIVES

- To practice prewriting technique
- To practice using sensory details, colorful words, and feeling words to describe characters and how they feel

Teach/Model Tell students that before they write about a character they should spend some time planning what vivid words and details they might include in their story. Read aloud the text at the top of the page. Direct students to read the example.

Guided Practice Read aloud the directions for Part A and the sentence about Keisha. Have students suggest sensory details, colorful words, and "feeling" words that might express Keisha's fear as she finds herself too far up in the tree. Guide students to complete Part A.

Independent Writing Practice After reading the directions for Part B, have students complete it independently. Then invite volunteers to read aloud their descriptions of how Keisha got down from the tree.

LESSON 4: REVIEW VOICE

Name ____________________

Practice with Writer's Craft

Review Voice

When you write a story, you create characters. You can use voice to show what the characters are like and how they feel. Read the example in the box below. It shows how one third grader began to use voice to create a character.

> **Felix has forgotten to pick up his little brother after preschool.**
> **Sensory Details:** dark streets, cold rain
> **Colorful Words:** terrible sinking feeling, tearstained face
> **"Feeling" Words:** worried, scared, relieved

A. Read the sentence below. Then think of words and details to add that can show how the character feels. Write your sentences on the lines.

Keisha climbed too far up in the tree.

1. Write a sensory detail to show how the tree bark might feel to Keisha.

   Possible response: The bark was rough under her hands.

2. Write a sentence with colorful words or "feeling" words to show that Keisha is afraid.

   Possible responses: Keisha clung to the tree, trembling; She tried to shout, but her throat closed up.

B. Write three or four sentences that describe how Keisha tries to get down from the tree. Include sensory details, colorful words, and "feeling" words. If you need more space, use another sheet of paper.

Responses will vary, but should include sensory details, colorful words, and words that express feelings.

Writer's Companion • UNIT 1
Lesson 4 *Review Voice*
28

Reaching All Learners

BELOW LEVEL

Have students who are writing below grade level answer specific questions to help guide their writing. Ask what Keisha's face might look like when she is scared. Invite them to share some things they might say if they were scared.

ADVANCED

Challenge students to write a paragraph about a character who is watching Keisha in the tree. Direct the students to use the same writing strategies to describe this character's voice as he or she watches Keisha in the tree.

ENGLISH LANGUAGE LEARNERS

Students who are learning English can write some colorful and "feeling" words in their native language. Once written, assist the student in translating the words into English. They can then work with a partner to write sentences using these words.

The Parts of a Story

OBJECTIVES
- To understand how a story is organized
- To analyze a Student Model

Teach/Model Tell students that showing a character's viewpoint is just one element that makes a good story. Read aloud, or have a volunteer read aloud, the introduction at the top of the page. Explain that the writing model on this page is the first draft of a student's story. Have students read aloud the steps for how to organize a story. Then read the model aloud. As you do, ask students to pay attention to how the details are organized.

Guided Practice Guide students to underline the sentence that introduces Felix.

Independent Writing Practice Have students complete Items 2–5 independently. Then discuss together the turning point of the story.

Use a Student Model

LESSON 4: REVIEW VOICE

Name ______

Focus on the Writing Form

The Parts of a Story

In this unit you have learned how writers use voice to show their own feelings and those of their characters. But a **story** has more than just a writer's voice. A story includes characters, a setting, and a plot. Below is a first draft of a story written by a third grader. As you read, think about how the student organized the story and also used voice to show how the characters feel. Then answer the questions.

Student Model

DRAFT

Felix Forgets
by Marcos

While waiting in the park, Felix fell asleep. Suddenly, he opened his eyes. He looked at his watch. It was five o'clock. He had a terrible sinking feeling. [He had forgotten to pick up Ramon after preschool!]

Felix sprang to his feet. He raced through the dark, wet streets. As he ran, he worried about his brother. Felix imagined his tear stained face. Ramon was only four. Ramon would never forgive him. Felix felt terribly guilty.

Felix burst through the door of the preschool. The teacher looked up from her desk and smiled. There was Ramon, happily playing with toy cars. He saw his brother and grinned. Felix melted at the sight. "Hey, Felix," Ramon shouted. "You are *late!*"

Introduce the setting, characters, and problem.

Organize events in the order they happened.

Provide details. Give sensory details that create mental pictures.

Create a strong ending. Explain how the main character solved the problem.

TIP: Use a mix of different types of sentences in a story. Include statements, exclamations, questions, and commands.

1. Underline the sentence that introduces Felix.
2. Draw [ ] around the sentence that tells Felix's problem.
3. Circle the sentences that tell how Felix feels about his problem.
4. In your own words, tell how Felix feels at the end of the story.

Possible response: Felix feels relieved and loving towards his brother.

29 Writer's Companion ■ UNIT 1
Lesson 4 *Review Voice*

WRITER'S GRAMMAR: VARYING TYPES OF SENTENCES

Read the Writer's Grammar aloud. Then ask volunteers to find an example of a statement and an exclamation in the Student Model. Explain that varying sentences makes a story more interesting and engaging to read. Different sentence types can express a variety of feelings and express a character's voice. Write examples of other sentence types on the board:

question: *Where is Felix?*

command: *Hurry up, Felix.*

Evaluate a Story

OBJECTIVES

- To understand what to look for when evaluating a story
- To see how a 4-point rubric is used to evaluate a piece of writing
- To find out why one student writing model received a score of 4 (excellent)

Teach/Model Tell students that some of the writing they do in school and on writing tests is evaluated using a special chart called a rubric. Point out the rubric on the following page. Explain that on the next two pages they will learn how writing is scored using a rubric.

Guided Practice Read the directions for Part A aloud. Then read aloud the Student Model. Have students listen (without looking at the story) and think about why the Student Model is a success. Then have students silently read the model and the teacher's comments that go with it. Explain that the comments show some of the things a teacher might look for when evaluating a story. Have students identify the comments that have to do with viewpoint, sensory details, and organization.

LESSON 4: REVIEW VOICE

Name ______________________________

Evaluating the Student Model

Evaluate a Story

A. **Two students were asked to write a story about a person who faces a problem. The story below got a score of 4. When using a 4-point rubric, a score of 4 means "excellent." Read the story and the teacher comments that go with it. Find out why this story is a success.**

Student Model

DRAFT

The Scary Dog
by Diana

Anna had always been afraid of dogs. It didn't matter if they were big or little. Even their wagging tails scared her. So when she saw the puppy on the sidewalk, she thought about crossing the street. Then she noticed it was shaking. Its head and tail were down. It was so thin that Anna could count its ribs.

> Nice beginning! You tell about the main character, the setting, and the problem here.

Anna stopped near the puppy. It had that "wet dog" smell. It wasn't too terrible, though. The puppy was the saddest-looking thing she had ever seen.

> This detail appeals to the sense of smell.

> These colorful and "feeling" words tell how Anna feels. Good!

"Hey, pup," she said softly. "Don't you have a home?"

The puppy looked up and wagged, just a little. Carefully Anna reached down and held out her hand. Her heart beat very hard. The puppy came nearer. Nervously, it touched its nose to Anna's hand. Then it licked her. The rough feel of its tongue made Anna smile.

> This detail appeals to the sense of touch.

"I guess you should come with me," she told the puppy. "How does a bath and dinner sound to you?"

The puppy wagged so hard it nearly fell over. Anna laughed. She knew she would never be afraid of dogs again.

> Your ending is strong. Anna's problem is solved!

Writer's Companion ▪ UNIT 1
Lesson 4 *Review Voice* 30

WRITER'S STRATEGY: SCORING A RUBRIC

Show students how to score a rubric, using the one provided on page 31. Have them check the box in each row that best describes the model they just read. Next, have them write the corresponding number of the three checked boxes on each line in the first column (*Voice*, *Sensory Details*, and *Organization*). Then have them add those three numbers together and divide the sum by three. Explain that this is the score they gave the model.

Evaluate a Story

OBJECTIVES

- To read a Student Model that received a score of 2 (needs improvement)
- To understand what makes a story get as score of 2 on a rubric
- To use a rubric to evaluate a story

Teach/Model Tell students that they will read a story that received a score of 2 on a 4-point rubric. Read the directions for Part B and Student Model aloud. Have students listen (without looking at the story) and think about why this story needs improvement.

Guided Practice Ask students to comment on any strengths and weaknesses they noticed as they listened to the story. Then have students read the Student Model and the teacher's comments that go with it. Guide students to summarize why the story needs improvement.

Independent Writing Practice Read the directions for Part C aloud and explain how to use the 4-point rubric at the bottom of the page. Ask students to complete it independently to show what score they would give the story.

Use a Student Model

LESSON 4: REVIEW VOICE

Name ______________________

Evaluating the Student Model

B. This paragraph got a score of 2. Why did it get a low score?

Student Model

DRAFT

The Test
by Jennifer

Callie knew something was wrong. The classroom was quite. All the students were at their desks. Mr. Boland stood at the front of the room. He looked very serious. Callie sat down quickly. At her desk. It was the big test. Mr. Boland passed out booklets. She had forgotten. What could she do? "Mr. Boland, I forgot we were having the test today," Callie said to the teacher.

Mr. Boland smiled. "You also forgot what I said. It isn't the kind of test you can study for. Don't worry, Callie. Just do your best."

Callie looked at the booklet on her desk. She felt a little better.

Your opening is interesting, but there are some spelling and sentence construction errors.

You have not told how Callie feels about forgetting.

Some details that appeal to the senses might be good here.

Your ending does not tell what happened to Callie.

C. What score would you give the student's story? Put a number on each line.

	4	3	2	1
Voice ______	☐ The writer uses many colorful and feeling words.	☐ The writer uses some colorful and feeling words.	☐ The writer uses few or no colorful and feeling words.	☐ The writer does not show the characters' feelings.
Sensory Details ______	☐ The writer uses sensory details.	☐ The writer uses some sensory details.	☐ The writer uses few sensory details.	☐ The writer uses no sensory details.
Organization ______	☐ There is a beginning, middle, and ending. Events are in order.	☐ There are characters and a problem. Some events are in order.	☐ The problem is unclear. Events may be out of order.	☐ The writer has used no organizing structure.

31

Writer's Companion ■ UNIT 1
Lesson 4 *Review Voice*

WRITER'S GRAMMAR: PUNCTUATING DIALOGUE

Point out the dialogue that is included in the story. Explain how quotation marks enclose the spoken words in a piece of writing. Show how a comma is used to punctuate the end of some dialogue in order to indicate the speaker.

Extended Writing/Test Prep

OBJECTIVES

- To apply the craft of voice in a longer piece of writing
- To select a topic for an extended piece of writing and begin the prewriting process
- To become familiar with the kinds of writing prompts that appear on writing assessments

Teach/Model Tell students that the next two pages will give them an opportunity to select a topic for a longer composition and follow the steps of the writing process as they complete it. Explain that they can choose from three possible writing activities. Go over the choices listed in Part A. Have each student select one.

Guided Practice Review the writing forms covered in the theme. Then have students reread the Student Model tips in the lesson where that writing form was taught:

- A Descriptive Paragraph: Lesson 1, Page 11
- A Journal/Diary Entry: Lesson 2, Page 17
- A Personal Narrative: Lesson 3, Page 23
- A Story: Lesson 4, Page 29

Read the directions for Part B aloud. Then have students begin the prewriting process by completing this part.

Use a Student Model

Name ______________________

Extended Writing/Test Prep

Extended Writing/Test Prep

On the last two pages of this lesson, you will use what you have learned to write a longer written work.

A. Read the three choices below. Put a star by the writing activity you would like to do.

1. Respond to a Writing Prompt

   Writing Situation: Everyone has faced a problem in life.

   Directions for Writing: Think about a problem facing someone.

   Now write about someone who is facing a problem. Use sensory details, colorful words, and "feeling" words to show how characters feel.

2. Choose one of the pieces of writing you started in this unit:
   - a descriptive paragraph (page 10)
   - a journal/diary entry (page 16)
   - a personal narrative (page 22)

   Expand your draft into a complete piece of writing. Use what you have learned about voice.

3. Choose a topic you would like to write about. You may write a journal/diary entry, a personal narrative, or a story. Use voice to help your writing come alive.

B. Use the space below and on the next page to plan your writing.

TOPIC: ______________________

____

WRITING FORM: ______________________

HOW WILL I ORGANIZE MY WRITING: ______________________

____

SHARING AND DISCUSSING

Have students form pairs or small groups and discuss their choices of topics. Invite each student to explain what he or she will write about, how the composition will be organized, and how he or she will apply viewpoint to improve the text. Encourage partners to provide feedback and ask questions to prompt each writer's thinking, creativity, and process.

Extended Writing/Test Prep

OBJECTIVES

- To use a graphic organizer to plan a piece of writing
- To use the steps of the writing process to complete a longer piece of writing independently

Teach/Model Read the directions to Part C aloud . Then explain to students that in Part C they will use a graphic organizer to help them plan their writing. Provide each student with a photocopy of the appropriate graphic organizer (see the back pages of this guide). Model how to complete a graphic organizer for a story.

Guided Practice Provide help as needed as students complete the graphic organizers. Remind students that the graphic organizer is a general guide to use as they write; they can add details to their writing later.

Independent Writing Practice **Draft, Revise, Publish** Have students write their compositions on a separate sheet of paper. After students have finished the first draft, ask them to revise their work. Remind them to think about organization and how effectively they express their voice. Ask students how they would like to publish their work. You might suggest that students read their work aloud or create a class anthology.

LESSON 4: REVIEW VOICE

Name ______________________

Extended Writing/Test Prep

C. In the space below, draw a graphic organizer that will help you plan your writing. Fill in the graphic organizer. Write more notes on the lines below.

Notes

____

____

____

____

____

____

____

____

D. Do your writing on another sheet of paper.

33 Writer's Companion ▪ UNIT 1 Lesson 4 *Review Voice*

WRITER'S STRATEGY: READ ALOUD TO REVISE

Explain to students that reading aloud a composition can help writers to revise their work. Tell them that by reading their work aloud, they can hear how their writing flows, find grammatical or spelling errors, and catch what might confuse a reader. Suggest to students that they read their first drafts aloud before they begin to make revisions.

Answering Multiple-Choice Questions

OBJECTIVES

- To understand that some writing tests contain multiple-choice questions
- To understand that some multiple-choice questions involve first reading a passage

Teach/Model Tell students that some writing tests include multiple-choice questions. Explain that this lesson will help them learn strategies for how to answer this type of question. Read aloud the directions for Part A and the test tip. Model reading all of the answer choices for Question 1. Model how to recite the poem in the boxed tip before choosing the correct answer. Then demonstrate how to mark the answer.

Guided Practice Ask a volunteer to read aloud Question 2. Direct all students to silently read the answer choices. Ask them to identify sentence 6 in the passage. Invite volunteers to read the sentence aloud using each punctuation mark: first as a question, another as a statement, and the third as an exclamation. Have the students decide which punctuation mark is most appropriate. Tell students to mark the correct answer next to Item 2 on the Answer Sheet.

Independent Writing Practice Have students complete the last question independently.

Use a Literature Model

LESSON 5: WRITING TEST PRACTICE

Name ______________________________

Writing Test Practice

Answering Multiple-Choice Questions

A. For some multiple-choice questions, you will have to read a passage with numbered sentences. Read the test tip. Then practice this type of question.

Read the passage. Then read each question and fill in the correct answer on your Answer Sheet.

Who Is There?

(1) It had gotten dark fast. (2) Delia walked quickly. (3) Then she heard a voice next to her ask, "Who." (4) Delia screamed. (5) She cried, "Who is there (6) Come out right now" (7) Again the voice said, "Who?" (8) Then wings flapped. (9) Delia saw an owl fly off into the woods.

1. What change, if any, should be made in sentence 5?
   - F Add a question mark after *there*
   - G Add a period after *there*
   - H Add an exclamation point after *there*
   - I Make no change
2. What change, if any, should be made in sentence 6?
   - A Add a question mark after *now*
   - B Add a comma after *now*
   - C Add an exclamation point after *now*
   - D Make no change
3. What change, if any, should be made in sentence 9?
   - F Change *woods.* to woods?
   - G Change *woods.* to woods!
   - H Insert a period after *owl*
   - I Make no change

Test Tip:
Use this rhyme to help you use the correct punctuation:

Would you like to be my friend?
A question mark is the way to end.

I'd like to be your friend indeed.
A period is what I need.

Answer all test questions on this Answer Sheet.

1. (F) (G) (H) (I)　　3. (F) (G) (H) (I)

2. (A) (B) (C) (D)

Writer's Companion ▪ UNIT 1
Lesson 5 *Writing Test Practice*
34

SHARING AND DISCUSSING

Have students work in groups of two or three to compare their responses for the last test question. Ask them to share the thinking process they each used to select the answer.

Answering Multiple-Choice Questions (cont.)

OBJECTIVES

- To practice completing multiple-choice questions that involve reading a passage in which sentences are numbered
- To practice the strategy of reading a passage in order to identify and correct errors

Teach/Model Explain that the items in the sample test ask them how to write a sentence correctly. Read aloud the directions for Part B and the test tip. Direct students to read the passage silently. Model how to read the four choices. Show students how to look for a complete sentence as defined by the test tip. Demonstrate how to select and mark the correct answer.

Guided Practice Ask a volunteer to read aloud Item 2. Ask students to identify the sentences that begin with a capital letter to narrow their choices. Have the students decide which end mark is most appropriate. Tell students to mark the correct answer next to Item 2 on the Answer Sheet.

Independent Writing Practice Have students complete Item 3 independently.

Name ______________________________

Writing Test Practice

B. **For some multiple-choice questions, you will have to read a passage and then decide the best way to correct a sentence. Read the test tip. Then practice answering this kind of question on the Answer Sheet below.**

The following is a rough draft of a student's personal narrative. It may contain errors. Look for errors as you read it.

In the Desert

(1) Every spring I visit my grandparents in San Diego. (2) we always go for a walk in the desert (3) This year all the flowers were in bloom. (4) It was very beautiful. (5) I saw something move as we walked? (6) It was a snake. (7) "Watch out!" my grandma cried. (8) "that's a rattlesnake" (9) We watched it slither away. (10) It was our best desert walk ever.

1. How should sentence 2 be correctly written?
   - A we always go for a walk in the desert.
   - B We always go for a walk in the desert
   - C We always go for a walk in the desert.
   - D we always go for a walk in the desert?
2. How should sentence 5 be correctly written?
   - A I saw something move as we walked?
   - B I saw something move as we walked.
   - C i saw something move as we walked.
   - D I saw something move as we walked,
3. How should sentence 8 be correctly written?
   - A "That's a rattlesnake!"
   - B "That's a rattlesnake?"
   - C "that's a rattlesnake?"
   - D "That's a rattlesnake"!

Test Tip: A complete sentence must begin with a capital letter and end with an end mark.

Answer all test questions on this Answer Sheet.

1. Ⓐ Ⓑ ● Ⓓ
2. Ⓐ ● Ⓒ Ⓓ
3. ● Ⓑ Ⓒ Ⓓ

ASSESSING STUDENT RESPONSES

If a student answers a test item incorrectly, try to find out why. An incorrect response may result from being unable to identify the error in the sentence or marking an answer in the wrong place on the Answer Sheet. Invite the student to explain the thinking behind the choice he or she made in answering the question. If the student simply mismarked the Answer Sheet, remind him or her to slow down when filling in the Answer Sheet. If the student was unable to recognize the correct response, make sure to direct him or her to reread the test tip. Review the rules for grammar and punctuation as necessary.

Answering Multiple-Choice Questions (cont.)

OBJECTIVES

- To become familiar with test items that require interpreting a graphic organizer
- To practice using an Answer Sheet

Teach/Model Explain that some test items require reading and interpreting a graphic organizer. Read aloud the directions for Part C. Have students look at the graphic organizer and think about the kind of writing Noah plans to do. Read aloud the test tip. Ask students what kind of writing would be written about the writer's own experience. Demonstrate how to select and mark the correct answer.

Guided Practice Ask a volunteer to read aloud Question 2. Guide students to answer the question. Ask students to think aloud to rule out each response and explain the correct choice. Invite students to mark the correct answer for Question 2.

Independent Writing Practice Have students complete Question 3 independently.

LESSON 5: WRITING TEST PRACTICE

Name ______________________

Writing Test Practice

C. For some writing tests, you will have to answer questions based on a graphic organizer. Follow the directions to practice answering this kind of question. Use the test tip to help you.

Noah made the plan below to organize ideas for a paper. Read his plan to answer questions 1–3.

Topic: My Entry into an Art Contest

	Beginning	Middle	End
What happened?	I painted a picture. I like to draw, too.	I entered the picture in an art contest at school.	I won second place. I got a silver ribbon.
How did I feel?	excited	scared, nervous	disappointed proud

1. Based on the information in Noah's writing plan, what is Noah planning to write?
   - Ⓐ a research report
   - Ⓑ a news article
   - Ⓒ a personal narrative
   - Ⓓ a how-to paragraph
2. Which detail is off topic and should be taken out of Noah's writing plan?
   - Ⓕ I entered the picture in an art contest.
   - Ⓖ I won second place.
   - Ⓗ I got a silver ribbon.
   - Ⓘ I like to draw, too.
3. Which detail should be placed in the box labeled "End"?
   - Ⓐ Alex Cody won first place.
   - Ⓑ The picture was of a lake.
   - Ⓒ The art contest was on Tuesday.
   - Ⓓ My class liked my painting.

Test Tip: Events in a writing plan should be in time order.

Writer's Companion ■ UNIT 1
Lesson 5 *Writing Test Practice* 36

Reaching All Learners

BELOW LEVEL

Students who are struggling to choose correct responses may benefit from working with a partner on practice items. Together, students can discuss how to rule out incorrect responses.

ADVANCED

Challenge students to explain the thinking behind the responses they provide. Have them write why an answer would be incorrect or correct.

ENGLISH LANGUAGE LEARNERS

English language learners may benefit from working through test items with a partner. Encourage them to read each item slowly. Have them eliminate the incorrect answers one by one.

Answering Multiple-Choice Questions (cont.)

OBJECTIVES

- To become familiar with test items that require reading, combining and ordering sentences
- To practice using an Answer Sheet

Teach/Model Explain that some test items require reading sentences and deciding how best to combine and order the sentences. Read aloud the directions for Part D. Read the test tip aloud and review the definition of *subject* and *predicate*. Then model how to read the two sentences in Item 1 and read the choices aloud. Explain how the sentences share the same predicate. Demonstrate how to select and mark the correct answer.

Guided Practice Ask a volunteer to read aloud Item 2. Guide students to answer the question. Ask students to think aloud to explain that the subject is shared by the sentences and to rule out each incorrect response. Tell students to mark the correct answer for Question 2.

Independent Writing Practice Have students complete Item 3 independently.

Use a Student Model

LESSON 5: WRITING TEST PRACTICE

Name ______________________

Writing Test Practice

D. **For some multiple-choice questions, you will have to read sentences and then decide the best way to combine and order them. Read the test tip and the questions. Then fill in the circle next to the correct answer.**

1. Combine the sentences in the box to make one sentence.

> Mollie takes dance lessons.
> Avram takes dance lessons.

Which sentence correctly combines the sentences in the box?

- Ⓐ Mollie takes and Avram takes dance lessons.
- Ⓑ Mollie and Avram take dance lessons. (filled)
- Ⓒ Mollie takes dance lessons and Avram takes dance lessons.

2. Combine the sentences in the box to make one sentence.

> The dog chased the ball.
> The dog caught the ball.

Which sentence correctly combines the sentences in the box?

- Ⓕ The dog and the dog chased and caught the ball.
- Ⓖ The dog chased the ball and the dog caught the ball.
- Ⓗ The dog chased and caught the ball. (filled)

3. Combine the sentences in the box to make one sentence.

> Lexie went to the store.
> Lexie bought a notebook.

Which sentence correctly combines the sentences in the box?

- Ⓐ Lexie went to the store and bought a notebook. (filled)
- Ⓑ Lexie went and she bought a notebook to the store.
- Ⓒ Lexie went to the store and Lexie bought a notebook.

Test Tip:

- You can combine sentences with the same subject by combining predicates.
- You can combine sentences with the same predicate by combining subjects.

37

Writer's Companion ■ UNIT 1
Lesson 5 *Writing Test Practice*

USING ACADEMIC LANGUAGE

Tests often ask students to choose a response that correctly does something. Explain to students that these words often mean that some responses may seem correct in some way. Warn students not to be distracted by choices that are correctly written but to only answer the actual question that is being asked.

Look at Time-Order Words and Sequencing

OBJECTIVES

- To understand the term *time-order words*
- To identify time-order words in literature

Teach/Model Tell children that good writers help readers understand the order of events in a story. Explain that readers may sometimes see clue words, such as *first, then, next,* and *last*, that help them understand the order of events. Elaborate by explaining that other clue words and phrases may tell time of day, days of the week, or dates. Make certain children understand that writers do not always use time-order clue words. Read aloud the introduction and passage from *Turtle Bay*. Have children listen for time-order words that help them understand when events occur and how the events are related in time to one another.

Guided Practice Read aloud the directions and Item 1 of Part B. Ask children to name the clue word that helps them tell what the children did after sweeping up litter. Then guide children in completing Part B.

Independent Writing Practice After reading aloud the directions, have students complete Part C independently.

Use a Literature Model

LESSON 1: TIME-ORDER WORDS AND SEQUENCING

Name ______________________________

Writer's Craft in Literature

Look at Time-Order Words and Sequencing

As you read, it is important to keep track of the order of the events. Writers use time-order words such as *now, first, next,* and *last* to tell us when things happen in a story.

A. Read the following Literature Model. Notice how the writer uses time-order words to tell you the order of events.

Literature Model

The three friends swept up all the litter dropped by the beachgoers during the day. Then they sat on the rocks and watched more turtles coming ashore. There were lots of them—all huge and old and wise—just like Jiro-San.

"Now," said Jiro-San, "you must be patient, and wait to hear from me again."

Eight weeks later, Jiro-San told the children to meet him at dusk.

—from *Turtle Bay*
by Saviour Pirotta

B. Identify the words that tell you the order of events.

1. Underline the sentence that tells you what the friends did *after* they swept up the litter.
2. Circle the time-order word in the sentence you underlined.

C. Write the answers to the questions on the lines.

1. Which words in the passage tell you how long it took to hear from Jiro-San again?
   Eight weeks later
2. At what time of day will Jiro-San meet the children?
   at dusk

Writer's Companion ■ UNIT 2
Lesson 1 *Time-Order Words and Sequencing* 38

WRITER'S STRATEGY: BRAINSTORMING

Tell children they can brainstorm to think of many ideas that will be helpful in writing. Have children brainstorm time-order clue words and phrases. Compile a class list children can use as they complete writing activities throughout this and other units.

Explore Time-Order Words and Sequencing

OBJECTIVES

- To explore and understand *time-order* and *sequence*
- To deepen students' understanding of how time-order words enhance writing
- To identify order of events in literature

Teach/Model Use the diagram at the top of the page to help children recognize specific time-order words they might see in literature. Include all the sequence words in the diagram as you tell a brief oral story. Then invite a volunteer to use the words to tell a new story.

Guided Practice Explain that writers use time-order words to "walk readers through" a scene or series of scenes. Write the example for Part A on the board. Read the directions aloud, pointing out how they apply to the example. Have a volunteer write and complete Item 1 on the board.

Independent Writing Practice Tell children to complete the rest of Part A on their own. After reading aloud both sets of directions, have children do Parts B and C independently.

LESSON 1: TIME-ORDER WORDS AND SEQUENCING

Name ______________________

A Closer Look at Writer's Craft

Explore Time-Order Words and Sequencing

Writers use words such as *before*, *next*, and *a day later* to show **time order** or **sequence**.

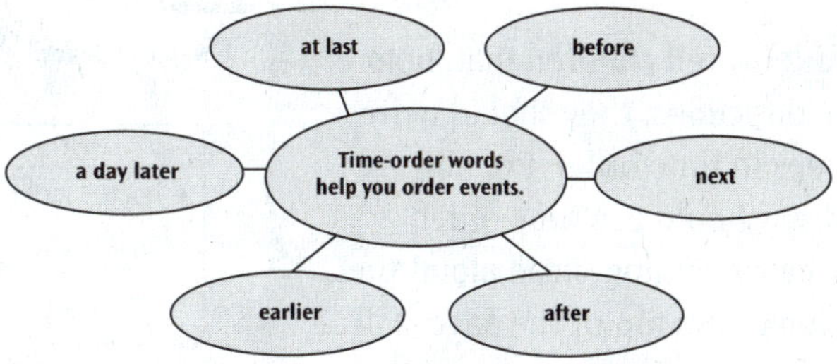

A. Underline the time-order word or words in each sentence. Then circle the event that happened first.

Example I always eat my toast after I eat my eggs.

1. Make sure to prepare all the ingredients before you start to cook.
2. First I heard a crack of thunder, and later I saw lightning.
3. I did my homework and then watched some TV.

B. Read these sentences from *Turtle Bay*, and underline the time-order word.

The boy and the old man swept the beach from one end to the other. They collected all the rubbish and put it in Jiro-San's cart. Soon the beach was cleaner than it had been all summer.

C. What was the second thing the boy and the old man did? Write your answer on the lines.

They collected the rubbish and put it in the cart.

39 Writer's Companion ■ UNIT 2
Lesson 1 *Time-Order Words and Sequencing*

WRITER'S STRATEGY: GRAPHIC ORGANIZERS

Explain that completing a graphic organizer can help children as they prepare to write. Point out the graphic organizer on the student page. Guide children in recognizing how it helps them prepare to set events out in time order. Ask how this graphic organizer could help them as they write. Create a variety of graphic organizers on the board, such as a word web, a flow chart, and a beginning-middle-end story map. Discuss how each type of graphic organizer can help children prepare to write.

Use Time-Order Words and Sequencing

OBJECTIVES

- To prepare to write directions by ordering the steps in a process
- To write directions that explain steps in a process

Teach/Model Tell children that, before they write directions, they should write out the steps in time order. Explain that children should use time-order words in their planning. Read aloud the introduction at the top of the page. Ask volunteers to read the four parts of the chart aloud.

Guided Practice Read aloud the directions for Part A. Then guide children to write the name of a snack they would like to make. If they need help, direct them to the name of the snack in the example. Then help them complete the chart.

Independent Writing Practice Have children complete Part B independently. Invite volunteers to read their paragraphs to the class.

LESSON 1: TIME-ORDER WORDS AND SEQUENCING

Name ______________________

Practice with Writer's Craft

Use Time-Order Words and Sequencing

Before you write directions, make sure you know the order of the steps. Here is how one student planned a recipe for her favorite snack.

Example

Name of Snack: Patty's Perfect Pizza

First	Then	Next	Finally
• toast muffin	• add sauce	• add cheese • sprinkle grated cheese	• press and pull apart

A. Think about a snack you like to make. Name the snack. Then fill out the chart.

Name of Snack: ______________________

First	Then	Next	Finally

B. Use information from your chart to write directions that explain how to make your snack. Use another sheet of paper, if you need to.

Answers will vary, but should include time-order words from the chart above.

____

____

Writer's Companion ■ UNIT 2
Lesson 1 *Time-Order Words and Sequencing* 40

Reaching All Learners

BELOW LEVEL

Have children divide a piece of paper into four squares, numbered *1, 2, 3,* and *4.* Guide children in drawing an illustration of the appropriate step in each square. Help them write a phrase to describe the drawing and then copy the phrase to the appropriate column in the chart.

ADVANCED

Encourage children writing above grade level to replace the clue word at the top of each chart column with a more interesting time-order word or phrase. Have them add sensory words to the description of each step.

ENGLISH LANGUAGE LEARNERS

Pair less fluent English speakers with more fluent English speakers. Have children act out the steps in the process as they tell what each step is. Then have children work together to write the appropriate English phrases in their charts.

The Parts of Directions

OBJECTIVES

- To understand the terms *directions* and *recipe*
- To understand how to organize directions
- To analyze a Student Model

Teach/Model Explain that the writing model on this page is a first draft of a recipe. Read aloud the introduction and the steps that tell how to organize this type of paragraph. Then read the model aloud. As children listen, have them pay careful attention to sequence words. Ask them to listen to determine if all the steps are written in an order that makes sense.

Guided Practice Point out the items below the Student Model. Read the first item aloud. Then guide children to identify and box the sentence that introduces the topic. Explain that this sentence tells the reader what the directions are for. Point out and read aloud the word *recipe* in Item 5. Make certain children understand that a recipe is a list of ingredients with directions that tell how to make something.

Independent Writing Practice Have children complete Items 2–5 independently.

Use a Student Model

LESSON 1: TIME-ORDER WORDS AND SEQUENCING

Name ______________________

Focus on the Writing Form

The Parts of Directions

Good **directions** give the steps of a task in time order. The paragraph below is a first draft of a recipe that was written by a third grader. As you read, think about how the student organized the steps. Then answer the questions.

Student Model

DRAFT

Patty's Perfect Pizza
by Patty

Here's a great way to make your own pizza. First, divide an English muffin in half and toast it. Put the two halves you just toasted on a plate. Spread a little tomato sauce on each half. Before you start, get together the ingredients. Next, unwrap a couple of pieces of sliced cheese, put the cheese slices on top of the sauce, and then sprinkle some grated cheese on top. Then, press the two halves together. You have to press for about five seconds. The cheese will have a chance to melt. At last, pull the halves apart carefully and put the halves back on the same plate you had at the beginning.

Introduce the topic with a sentence that clearly states the task.

Organize the steps in time order using sequence words.

Include details readers need to know to complete the task.

1. Which sentence introduces the topic? Put a box around it.
2. Underline the sentence that tells what to do last.
3. Which sentence seems to be out of order? Circle it. Then draw an arrow to the place where you think this sentence belongs.
4. Find the time-order words. Write them below.

   Before you start, First, Next, Then, At last

5. How do time-order words help you with the recipe?

   They tell the order of the steps.

Writer's Grammar
You can join two complete sentences to make a compound sentence. Join the sentences with a comma followed by *and* or *but*. "I ran home, and then I came back," is a compound sentence. Find a compound sentence in the Student Model.

Writer's Companion ■ UNIT 2
41 Lesson 1 *Time-Order Words and Sequencing*

WRITER'S GRAMMAR: COMPOUND SENTENCES

Tell children that combining short sentences can make writing clearer, more interesting, and less choppy. Explain that when two sentences relate, children can combine them with a comma and the word *and*. Write this example on the board:

I finished my paper. + I put it on the desk. = I finished my paper, and I put it on the desk.

Ask children to offer more examples and to join them. Then have children find a compound sentence in the model.

Evaluate Directions

OBJECTIVES

- To understand how to evaluate directions
- To use a checklist to evaluate directions
- To suggest improvements for directions

Teach/Model Discuss why it is important to evaluate a piece of writing. Explain that even the best writers usually write more than one draft. This allows them to note things done well and improve things that could be done better. Use the points at the top of the page to discuss what to look for when evaluating directions.

Guided Practice Read aloud the directions for Part A. Guide children in writing responses as they refer to the Student Model for information. Emphasize the value of using a checklist to evaluate a piece of writing. Explain that the checklist encourages writers to look at specific parts of the writing. Point out the checklist in Part B. Explain that it contains many of the writing skills children have practiced in this lesson.

Independent Writing Practice After reading aloud both sets of directions, have children complete Parts B and C independently.

LESSON 1: TIME-ORDER WORDS AND SEQUENCING

Name ______________________

Evaluating the Student Model

Evaluate Directions

When you evaluate directions, ask yourself these questions:

- Does the writer put the steps in the correct sequence? (Look for time-order words such as *before, first, next,* and *finally*.)
- Does the writer include clear, easy to follow directions? (Look for details that give important information, such as how long the halves should be pressed together.)

A. Reread the Student Model on page 41. Then answer these questions.

1. Write two sentences from the model that contain time-order words. Circle the time-order words.

   Possible response: Before you start, get together the ingredients. Then, press the two halves together.

2. What happens after the grated cheese is added?

   Possible response: The two halves are pressed together.

B. Now evaluate the Student Model. Put a check in the box next to each thing the writer has done well. If you do not think the writer did a good job with something, do not check the box.

- ☐ The first sentence clearly states the task.
- ☐ The materials are listed in order near the beginning.
- ☐ The writer used a straightforward tone.
- ☐ The writer used details that give important information.
- ☐ The writer put sentences in sequence order.
- ☐ The writer used time-order words to help sequence the steps.

C. How do you think the writer could make the directions better?

Possible response: The materials list belongs at the beginning. One sentence needs to be moved. An unimportant detail should be deleted.

Writer's Companion • UNIT 2
Lesson 1 *Time-Order Words and Sequencing* 42

See the rubric on page 207 for another way to evaluate the Student Model.

EXTENDING THE CONCEPT: WORD CHOICE

Point out the phrases *a little tomato sauce, a couple of pieces,* and *some grated cheese*, in the Student Model. Explain that these phrases are not clear and do not tell the reader exactly what the writer means. Guide children in brainstorming more exact words to tell the reader how to make the pizza.

Revise by Deleting

OBJECTIVES

- To revise directions by deleting unimportant details
- To improve an earlier piece of writing by deleting unimportant details

Teach/Model Tell children that revising means finding ways to make writing clearer and more interesting. Explain that writers can make many kinds of changes. Explain that in this activity children will revise their writing by removing unimportant words.

Guided Practice Read aloud the introduction and the example. Make certain children understand that *deleting* is the same as removing. Discuss how deleting the words in the example makes information clearer. Guide children in revising Item 1 in Part A. Invite them to share their revisions.

Independent Writing Practice Have children complete the rest of Part A and then Part B independently. Then go on to the Sharing and Discussing activity.

LESSON 1: TIME-ORDER WORDS AND SEQUENCING

Name ______________________

Revising the Student Model

Revise by Deleting

One thing the student writer could have done better is to delete unnecessary words and details. Here is an example of how a sentence from the Student Model can be improved.

Example **Original:** At last, pull the halves apart carefully and put the halves back on the same plate you had at the beginning.

Revision: At last, pull the halves apart carefully and put them back on the plate.

A. Revise these sentences from the Student Model. Delete unimportant details.

1. Put the two halves you just toasted on a plate. Answers will vary. Possible responses are given.

   Put the two halves on a plate.

2. Next, unwrap a couple of pieces of sliced cheese, put the cheese slices on top of the sauce, and then sprinkle some grated cheese on top.

   Next, put the cheese slices on top of the sauce and sprinkle some grated cheese over everything.

3. Then, press the two halves together. You have to press for about five seconds.

   Then, press the two halves together for about five seconds.

B. Revise two of the sentences you wrote on page 40. Delete unnecessary words and details. Use more paper if necessary.

Answers will vary. Students should delete unnecessary words and details.

43 Writer's Companion ■ UNIT 2
Lesson 1 *Time-Order Words and Sequencing*

SHARING AND DISCUSSING

Have children read aloud their revised sentences from Part B. Invite them to discuss why they chose to revise as they did. Ask how the writer's voice came through in the revision. Invite volunteers to write their revisions on the board. Have children comment as to how conventions such as spelling, grammar, and usage were handled well. Guide children in kindly offering constructive criticism for conventions that need correction.

Look at Topic Sentence and Details

OBJECTIVES

- To understand the terms *topic sentence* and *details*
- To identify a topic sentence and details in literature

Teach/Model Explain that a topic sentence tells the main idea in a paragraph. Tell children that details give smaller bits of information about the main idea. Read aloud the directions for Part A and the passage from "Balto, the Dog Who Saved Nome". Have children listen for the topic sentence that gives the main idea. Tell them to also listen for details, information that tells more about the main idea.

Guided Practice Ask children to identify the topic sentence and tell the main idea in the paragraph. Then read the directions for Part B aloud and guide children in completing Part B. Be sure to point out details that support the main idea.

Independent Writing Practice After reading aloud the directions, have children complete Part C independently. Tell them to look for words that helped them figure out the main idea in the paragraph.

Use a Literature Model

LESSON 2: TOPIC SENTENCE AND DETAILS

Name ______________________________

Writer's Craft in Literature

Look at Topic Sentence and Details

The **topic sentence** of a paragraph tells you the main idea. The sentences that follow give examples with **details** that support the main idea.

A. Read the following passage. Notice how the writer uses a topic sentence and supporting details to give you information about a subject.

Literature Model

The first part of the trail to Nome led across sea ice. The ice wasn't anything like ice on a small pond or lake. It seemed much more *alive.* And no wonder. The water *under* the ice was moving up and down because of the storm. So the ice was moving up and down too. Up and down, up and down it went, like a roller coaster.

—from "Balto, the Dog Who Saved Nome," *Seven True Dog Stories* by Margaret Davidson

B. Identify the words that tell you the topic sentence and details.

1. Underline the topic sentence.
2. Circle one sentence that gives information about the main idea. Answers will vary. Possible response given.

C. Which words in the topic sentence tell you what the rest of the passage is about? Write them below.

across sea ice

Writer's Companion • UNIT 2
Lesson 2 *Topic Sentence and Details*
44

EXTENDING THE CONCEPT: IDEAS

Explain that supporting details are made up of clear examples. Tell children that these examples should include interesting, important information. Point out that writers should try to include the kinds of details that might surprise readers and grab their attention. Direct children to the Literature Model. Point out the ideas that the ice was *alive* and moving *like a roller coaster.* Ask how these ideas are unusual and how they support the main idea. Invite children to add additional unusual ideas to support the paragraph's topic.

Explore Topic Sentence and Details

OBJECTIVES

- To explore and deepen children's understanding of main idea in a topic sentence and details that support the main idea
- To determine which details do and do not support a main idea

Teach/Model Use the diagram at the top of the page to help children recognize that details support a topic sentence by telling more about it. Make certain children understand that *to support* means to hold up or explain further.

Guided Practice Explain that offering details makes the main idea stronger and clearer. Read aloud the directions for Part A. Have a volunteer read the example aloud. Point out that the detail about loving Charlie does not tell more about the main idea: how dogs help people. Guide students in completing Item 1 in Part A by having one volunteer identify the main idea and another identify a detail that does not belong. Make certain the volunteer explains why the detail does not belong.

Independent Writing Practice Invite children to complete the rest of Part A on their own. Then after reading aloud both sets of directions, have children do Parts B and C independently.

LESSON 2: TOPIC SENTENCE AND DETAILS

Name ______________________________

A Closer Look at Writer's Craft

Explore Topic Sentence and Details

Writers use a **topic sentence** to present the main idea. They use **details** to give examples and information about the main idea.

A **topic sentence** states the main idea.

Details can be examples. | **Details** support the main idea. | **Details** can be information.

A. **Underline the topic sentence, or main idea. Circle the sentence that does not support the topic.**

Example Dogs help people in many ways. Some work as seeing-eye dogs. Some, like sled dogs, pull heavy loads. Other dogs help people in wheel chairs. I love my dog Charlie very much.

1. Fall is my favorite season. The leaves turn colors. The air turns brisk. Summer is my next-favorite season. Fall air smells like winter is on the way.

2. The storm was very scary. First, lightning lit up the sky. Then, after about five seconds of silence, I heard a low rumbling. Heavy rain continued as more lightning struck. The red truck slowed down on the highway.

B. **Read these sentences from "Balto, the Dog Who Saved Nome." Underline the topic sentence. Circle three details that support the topic.**

The teams were ready. The first team pushed north through the storm to a little town. There a second team was waiting. It went on to another small town where a third team was ready to take the medicine farther.

C. **Suppose you are writing "Balto, the Dog Who Saved Nome." Add another detail to the paragraph above. Make sure it supports the topic sentence.**

____

____

45 Writer's Companion • UNIT 2
Lesson 2 *Topic Sentence and Details*

EXTENDING THE CONCEPT: ORGANIZATION

Point out that the paragraph in the Literature Model on page 44 and the one in Part A on page 45 are organized by main idea and details. Explain that Item 2 in Part A, even though it states a main idea and details, is organized by time order. Write *Main Idea and Details* and *Time Order* on the board. Ask children to name other ways to organize writing. Elicit these responses: Cause and Effect, Compare/Contrast, and Problem/Solution. Add these types of organization to the list. Ask children to discuss each type of organization and to give an example of a type of writing in which each would be most effective.

Use Topic Sentence and Details

OBJECTIVES

- To prepare to write by generating a topic sentence and supporting details
- To write a paragraph of information

Teach/Model Tell children that, before they write a paragraph of information, they should plan by writing a topic sentence and a list of supporting details. Explain that this will help them find out if they have a clear main idea and enough details to support it. It will also help them make certain that all their details relate to the main idea. Read aloud the introduction at the top of the page. Ask volunteers to read the topic sentence and details. Discuss how each detail supports the topic sentence.

Guided Practice Read aloud the directions for Part A. Then guide children to choose a pet and write a topic sentence. Guide them in writing supporting details.

Independent Writing Practice Have children complete Part B independently. Invite volunteers to share their choices of pet from the chart in Part A and their sentences from Part B.

LESSON 2: TOPIC SENTENCE AND DETAILS

Name ____________________

Practice with Writer's Craft

Use Topic Sentence and Details

Before you write a paragraph of information, make sure you have a topic sentence and enough details to support it. Here is how one student planned a paragraph of information about what makes a good pet.

Example **My Pet** Cat

Topic Sentence	Detail	Detail	Detail
• I think cats are the best pets to have.	• do not have to be walked	• mysterious	• fun to watch

A. Think about a pet you would like to have. Write what animal it is on the line. Then fill out the chart.

My Pet: Answers will vary.

Topic Sentence	Detail	Detail	Detail

B. Use information from your chart. Write a paragraph to explain why the animal you chose would make a good pet. Use more paper if you need it.

Answers will vary but should include details from the chart above.

Writer's Companion • UNIT 2
Lesson 2 *Topic Sentence and Details* 46

Reaching All Learners

BELOW LEVEL

If a child has difficulty generating a topic sentence, ask what his or her paragraph will be about. Guide the child in creating and understanding a word web.

ADVANCED

Encourage children writing above grade level to extend their planning charts to include additional details. Suggest that they include figurative language in the sentences they write.

ENGLISH LANGUAGE LEARNERS

Have pairs work to write the main idea on a long piece of paper and one detail on several self-stick notes below it. Then have children transfer the main idea to the topic sentence chart section and words from each note to a detail section.

The Parts of a Paragraph of Information

OBJECTIVES

- To understand how a paragraph of information is organized
- To analyze a Student Model

Teach/Model Read the introduction aloud. Explain that the writing model on this page is a first draft of a third grader's paragraph of information. First, read aloud the steps that explain how to organize this type of paragraph. Then read the model aloud. Have children listen for the main idea and supporting details. Ask them to also listen to figure out whether or not all the details tell about the main idea.

Guided Practice Point out the questions below the Student Model. Read the first question aloud and guide children to identify and underline the topic sentence. Remind them that the topic sentence tells the main idea.

Independent Writing Practice Have children complete Questions 2–3 independently. Make certain children understand that the word *unnecessary* in Question 3 means "not needed."

Use a Student Model

LESSON 2: TOPIC SENTENCE AND DETAILS

Name ______________________

Focus on the Writing Form

The Parts of a Paragraph of Information

A **paragraph of information** gives a topic sentence followed by details or examples to support it. Below is a draft of a paragraph written by a third grader. As you read, think about how the student organized her paragraph. Then answer the questions.

Student Model

DRAFT

Cats Are Cool
by Jessie

I think cats are the best pets to have. They are quiet and cute. They sleep on your bed. Also, they are fun to watch, and I love to watch them. Buster is two years old, and I got him at a shelter. Cats are very mysterious, but some people say dogs are too. They will do crazy things for no reason at all. They don't need to be walked, but they do have to have food and water. For instance, they will suddenly leap up and twirl around. They will stare for hours out the kitchen window or the bedroom window. You never know what cats are thinking. I love my cat Buster very much. If you get a cat you will not be sorry.

Introduce the topic by writing a sentence that clearly states the main idea of the pararagraph.

Organize the supporting details by deciding on an order that makes sense.

Develop your idea by writing sentences that present the details in the order you chose.

Conclude by writing a last sentence that restates the main idea.

1. What is the topic sentence of the paragraph? Underline it.
2. Which sentence repeats the main idea of the paragraph? Put a box around it.
3. Which sentence seems unnecessary? Circle it.

Writer's Grammar
A **common noun** names a person, place, or thing. A **proper noun** names a particular person, place, or thing. Make sure you begin each proper noun with a capital letter. Find a common noun and a proper noun in the model.

47 Writer's Companion ■ UNIT 2
Lesson 2 *Topic Sentence and Details*

WRITER'S GRAMMAR: COMMON AND PROPER NOUNS

Remind children that a noun names a person, place, or thing. Explain that a common noun names any person, place, or thing, but a proper noun names a particular person, place, or thing. Create two columns on the board, one for common nouns and one for proper nouns. Then write these example pairs in the columns: *girl/Lin; street/King Street; cat/Fluffy; dog/Buster.* Invite children to add new pairs. Point out how each proper noun begins with a capital letter.

Evaluate a Paragraph of Information

OBJECTIVES

- To understand how to evaluate a paragraph of information
- To use a checklist to evaluate a paragraph of information
- To suggest improvements for a paragraph of information

Teach/Model Discuss why it is important to evaluate writing. Explain that looking back at a piece of writing can help a writer discover another way to express ideas and approach organization. Use the points at the top of the page to discuss what to look for in evaluating a paragraph of information.

Guided Practice Read aloud the directions for Part A. Guide children in writing responses as they refer to the Student Model (page 47) for information. Explain that a checklist helps writers break down the paragraph and take a close look at each part. Point out the checklist in Part B. Encourage children by telling them that they have the knowledge they need to complete the checklist.

Independent Writing Practice After reading aloud both sets of directions, have students complete Parts B and C independently.

LESSON 2: TOPIC SENTENCE AND DETAILS

Name ______________________

Evaluating the Student Model

Evaluate a Paragraph of Information

When you evaluate a paragraph of information, ask yourself these questions:

- Does the paragraph have a clear topic sentence? (Look for a topic sentence that clearly states the main idea.)
- Does the writer include examples supporting the main idea? (Look for details that give information about the topic.)

A. Reread the Student Model on page 47. Then follow the directions below.

1. Name two details that tell why cats make the best pets.

   Possible responses: They sleep on your bed. They are fun to watch.

2. Give an example of a detail that supports the idea that cats are mysterious.

   Possible response: They will do crazy things for no reason at all.

B. Now evaluate the Student Model. Put a check in the box next to each thing the writer has done well. If you do not think the writer did a good job with something, do not check the box.

- ☐ The writer included a topic sentence that clearly states the main idea.
- ☐ The detail sentences give examples that support the main idea.
- ☐ The writer put sentences in an order that makes sense.
- ☐ The conclusion restates the main idea of the paragraph.

C. How do you think the writer could make the paragraph better?

Possible response: The writer should delete unimportant details such as "Buster is two years old, and I got him at a shelter," and "They don't need to be walked, but they do have to have food and water."

Writer's Companion ▪ UNIT 2
Lesson 2 *Topic Sentence and Details* 48

See the rubric on page 207 for another way to evaluate the Student Model.

WRITER'S STRATEGY: PEER EDITING

Explain that "peer editing" means telling a classmate what has been done well and what can be improved. Tell children they should always exercise respect when editing the work of a peer, another writer. Point out that they should begin by telling what has been done well. Then they can continue by gently and respectfully telling what could be done better. Explain that a peer editor should always point out a specific example in the writing for each point, along with a specific suggestion as to how it can be improved.

Revise by Deleting

OBJECTIVES

- To revise a paragraph of information by deleting unnecessary or unimportant details
- To revise a paragraph of information by combining sentences
- To improve an earlier piece of writing by deleting unimportant details

Teach/Model Tell children that writers revise to make their writing better and clearer. Explain that looking back at a completed piece of writing can help writers see what needs to be improved. Elaborate by telling children that in this lesson they will be revising by deleting words and phrases that are not important or do not tell more about the topic.

Guided Practice Read aloud the introduction and example. Have children return to the model on page 47. Guide them in recognizing that the detail about dogs does not tell more about the topic—cats. Continue to guide children as they revise Item 1 in Part A. Invite suggestions for alternate ways to revise.

Independent Writing Practice Have children complete the rest of Part A and Part B independently. Then invite them to participate in the Sharing and Discussing activity below.

LESSON 2: TOPIC SENTENCE AND DETAILS

Name ______________________

Revising the Student Model

Revise by Deleting

One thing the student writer could have done better is to delete details that do not fit or are not important. Here is an example of how a sentence from the Student Model can be improved by deleting unnecessary details.

Example **Original:** Cats are very mysterious, but some people say dogs are too.

Revision: Cats are very mysterious.

A. **Revise these sentences from the Student Model. Delete unimportant details and combine sentences when necessary.** Answers will vary. Possible responses are given.

1. Also, they are fun to watch, and I love to watch them.

   Also, they are fun to watch.

2. They will stare for hours out the kitchen window or the bedroom window.

   They will stare for hours out the window.

3. They are quiet and cute. They sleep on your bed.

   They are quiet, cute and will sleep on your bed.

B. **Revise the paragraph you wrote on page 46. Delete any words that are unimportant or do not support the topic. Continue on another sheet of paper.**

Answers will vary. Students should delete or revise extraneous details.

SHARING AND DISCUSSING

Have children read their revised sentences to the class. Ask them to tell why they decided to delete certain details. Guide children in discussing why the details did not relate to the main idea. Then invite them to tell a new detail that would support the main idea. Ask children to discuss how they will make sure all details support the main idea in future pieces they write.

Look at More About Topic Sentence and Details

OBJECTIVES

- To understand the terms *main idea* and *details* in greater depth
- To identify a topic sentence and details in literature
- To write a detail that supports a main idea

Teach/Model Remind children that a topic sentence tells the main idea in a paragraph. Explain that all details should support the main idea. These details should tell important information about the topic. Have children listen for the main idea and details as you read aloud the paragraph from "Wild Shots, They're My Life".

Guided Practice Read aloud the directions for Part B. Guide children through Items 1 and 2, having them identify the main idea and the topic sentence. Then guide them through Item 3 as they box a detail that supports the main idea.

Independent Writing Practice Have children complete Part C independently. Remind them that the detail should support the main idea.

Use a Literature Model

LESSON 3: MORE ABOUT TOPIC SENTENCE AND DETAILS

Name ______________________________

Writer's Craft in Literature

Look at More About Topic Sentence and Details

The topic sentence of a paragraph tells you the main idea. Other sentences in a paragraph usually give details that support the main idea.

A. Read the following paragraph. As you read, look for the main idea and the details.

Literature Model

The Galapagos Islands are famous for their huge tortoises. In fact, *galapagos* means "tortoises" in Spanish. When early Spanish-speaking explorers came to the islands, they saw *tons* of these big fellas.

—from "Wild Shots, They're My Life" *Ranger Rick* Magazine by Tui De Roy

B. Identify the topic sentence and details.

1. Underline the topic sentence.
2. Circle the main idea in the topic sentence.
3. Put a box around a detail that tells about the main idea. See Possible responses

C. Write another detail.

Details will vary. It should be any detail about tortoises.

____

____

Writer's Companion • UNIT 2
Lesson 3 *More About Topic Sentence and Details* 50

SHARING AND DISCUSSING

Invite children to share the details they wrote for Part C. Ask why they wrote the details they did, how the details tell more about the paragraph, and how they could make the details more interesting. Have small groups participate in a group to combine their details in a clear and interesting way to add to the paragraph. Invite children to share their new paragraphs by first writing them and then reading them aloud to the class.

Explore More About Topic Sentence and Details

OBJECTIVES

- To explore and deepen children's understanding of topic sentences and supporting details
- To identify main ideas, topic sentences, and supporting details
- To write details

Teach/Model Use the diagram at the top of the page to reinforce children's understanding that the main idea appears in the topic sentence. Point out how the details provide more information about the main idea.

Guided Practice Remind children that details in a paragraph should relate to the main idea. Explain that it is important to determine the main idea in the paragraph and then to look for details. Read aloud the directions and example for Part A. Make certain children understand that the topic is *a great snack* and the example of a great snack is *peanut butter* on a *banana*. This example is the detail. Guide children in completing Item 1 in Part A.

Independent Writing Practice Tell children to complete the rest of Part A and all of Parts B and C independently.

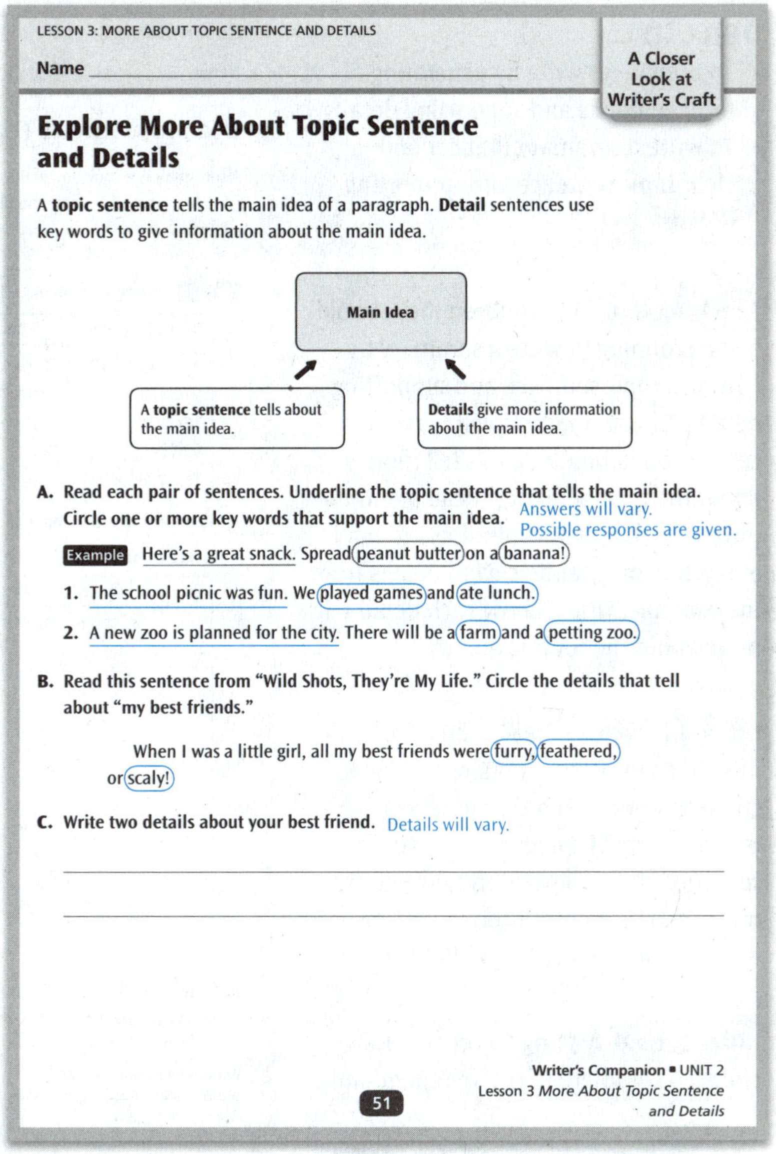
LESSON 3: MORE ABOUT TOPIC SENTENCE AND DETAILS

Name ________________________

A Closer Look at Writer's Craft

Explore More About Topic Sentence and Details

A **topic sentence** tells the main idea of a paragraph. **Detail** sentences use key words to give information about the main idea.

A. **Read each pair of sentences. Underline the topic sentence that tells the main idea. Circle one or more key words that support the main idea.** Answers will vary. Possible responses are given.

Example Here's a great snack. Spread peanut butter on a banana!

1. The school picnic was fun. We played games and ate lunch.
2. A new zoo is planned for the city. There will be a farm and a petting zoo.

B. **Read this sentence from "Wild Shots, They're My Life." Circle the details that tell about "my best friends."**

When I was a little girl, all my best friends were furry, feathered, or scaly!

C. **Write two details about your best friend.** Details will vary.

____

____

51

Writer's Companion ▪ UNIT 2
Lesson 3 *More About Topic Sentence and Details*

WRITER'S GRAMMAR: EXCLAMATIONS

Write these sentences from Parts A and B on the board:

- *Spread peanut butter on a banana!*
- *When I was a little girl, all my best friends were furry, feathered, or scaly!*

Point out the exclamation mark at the end of each sentence. Explain that the exclamation mark helps the writer show strong feeling. Ask children to tell the strong feeling they think the writer is trying to tell in each sentence. Then invite volunteers to write new exclamations on the board. Have them name the strong feeling they are trying to convey to the reader.

Use More About Topic Sentence and Details

OBJECTIVES

- To prepare to write by generating a topic sentence and supporting details
- To write a summary that contains a clear topic sentence and supporting details

Teach/Model Tell children they should begin planning to write a summary by writing a topic sentence and supporting details. Explain that this will make certain that all their details tell more about the topic sentence. Read aloud the introduction and ask volunteers to read aloud the topic sentence and details from the example. Discuss how each detail tells more about the topic sentence.

Guided Practice Read aloud the directions for Part A, making certain children understand that *observed* means "seen." Explain that the topic sentence children write should tell the main idea. Guide children in writing details that support the main idea.

Independent Writing Practice Have children complete Part B independently.

LESSON 3: MORE ABOUT TOPIC SENTENCE AND DETAILS

Name ______________________

Practice with Writer's Craft

Use More About Topic Sentence and Details

A summary is a short form of a longer piece of writing. A summary includes a main idea and the most important details. When you write a summary, use your own words. Here is how one third grader planned her summary.

Example **Name of Animal:** Tortoise

Topic Sentence	Detail	Detail	Detail	Detail
The writer was nose to nose with a tortoise.	only a camera between them	had to move out of the way	eating plants	bent down to take the picture

A. **Think about an animal you have observed. Name the animal and fill out the chart.**

Name of Animal: Answers will vary.

Topic Sentence	Detail	Detail	Detail	Detail

B. **Write a summary of the most important information from your chart. Include a topic sentence and details. Use another sheet of paper for your writing.**

Answers will vary, but should include details from the chart above.

Writer's Companion ▪ UNIT 2
Lesson 3 *More About Topic Sentence and Details*
52

Reaching All Learners

BELOW LEVEL

If a child has difficulty naming an animal or thinking of a topic sentence, invite the child to participate in a brainstorming activity. Prompt with questions such as: "Do you have a pet? Have you been to the zoo? Have you read a story or an article about an animal?"

ADVANCED

Encourage children writing above grade level to think of a very unusual animal or to imagine what an animal from another planet or deep in the ocean might be like. Have children complete their charts with a topic sentence and details that relate to the animal.

ENGLISH LANGUAGE LEARNERS

Provide books with animal illustrations. Pair English language learners with proficient English speakers. Have pairs use the books to discuss vocabulary related to specific details about one of the animals. Then have pairs work together to complete the chart and paragraph.

The Parts of a Summary

OBJECTIVES

- To understand how a summary is organized
- To analyze a Student Model

Teach/Model Read the introduction aloud. Explain that the writing model on this page is a first draft of a third grader's summary. Read aloud the steps that explain how to organize this type of paragraph. Emphasize that the summary should include only the most important details and should not include the writer's opinion. Make certain children understand that an opinion is a thought, feeling, or belief that cannot be proven true or false. Provide these examples: *Fact: Our room is in our school building. Opinion: Our room is beautiful*. Then read the model aloud. Have children listen for the main idea, details, and any opinions that might appear in the paragraph.

Guided Practice Point out the items below the Student Model. Read aloud the first question and guide children to identify and underline the main idea.

Independent Writing Practice Have children complete Items 2–4 independently.

LESSON 3: MORE ABOUT TOPIC SENTENCE AND DETAILS

Name ______________________________

Focus on the Writing Form

The Parts of a Summary

When you write a **summary** from a longer piece of writing, always begin with a topic sentence. Then include only the most important ideas. Below is the first draft of a summary written by a third grader. As you read, think about how the student organized the summary. Then answer the questions.

Student Model

DRAFT

Summary
by Rachel

The writer was nose to nose with a big tortoise. There was just a camera between them. The tortoise didn't notice her there. She had to move out of the animal's way before it bumped right into her! It must have been so scary. The tortoise was busy eating tasty plants. Tortoises don't mind the spines of a cactus. Then the writer bent down and snapped the picture. The picture was great. The writer said getting close is the best way to get great action shots. CHOMP!

Write a topic sentence that tells the main idea.

Choose the most important ideas or details from the longer selection. Restate them in your own words.

Leave out less important opinions and details.

Put your **ideas in an order** that makes sense. Follow the order of the ideas in the selection you are summarizing.

Responses for 2–4 will vary.

1. What sentence tells the main idea of the summary? Underline it.
2. Which sentence gives an important idea? Put a box around it.
3. Circle a detail.
4. Put [ ] around an opinion.

Writer's Grammar
Most of the time you add *–ed* to a verb (action word) to change the action to the past: *jump–jumped*. Other times, the past form of a verb changes in a different, or irregular, way: *send–sent*. Find past forms of verbs in the Student Model.

53

Writer's Companion ▪ UNIT 2
Lesson 3 *More About Topic Sentence and Details*

WRITER'S GRAMMAR: IRREGULAR VERBS

Remind children that verbs name an action. Explain that children can add *–ed* to many verbs to put the action in the past, but some verbs do not follow that rule. On the board, write the following:

Present	Past
run	ran
am	was
dig	dug
make	made

Evaluate a Summary

OBJECTIVES

- To understand how to evaluate a summary
- To use a checklist to evaluate a summary
- To suggest improvements for a summary

Teach/Model Discuss why it is important to evaluate writing. Explain that evaluating helps writers tell if all the parts of the writing fit together well and express ideas clearly. Use the points at the top of the page to discuss what to look for when evaluating a summary.

Guided Practice Read aloud the directions for Part A. Guide children in writing responses as they refer to the Student Model for information. Explain that a checklist helps children focus on specific parts of the writing. Point out the checklist in Part B. Remind children that they have been working with these skills throughout the lesson, so they are prepared to complete the checklist.

Independent Writing Practice After reading aloud both sets of directions, have children complete Parts B and C independently. Remind them to look back at the model as they complete Part C.

LESSON 3: MORE ABOUT TOPIC SENTENCE AND DETAILS

Name ______________________

Evaluating the Student Model

Evaluate a Summary

When you evaluate a summary, ask yourself these questions:

- Does the summary begin with a sentence that tells the main idea? (Look for a sentence that clearly states the main idea.)
- Does the summary include only the most important ideas? (Make sure there are no opinions or details.)

A. Reread the Student Model on page 53. Then follow the directions below.

1. Name two important ideas from the summary.

   Possible responses: There was just a camera between them. The writer said getting close is the best way to get great action shots.

2. Give two unimportant details from the summary.

   Possible responses: Tortoises don't mind the spines of a cactus. The picture was great.

B. Now evaluate the Student Model. Put a check in the box next to each thing the writer has done well. If you do not think the writer did a good job with something, do not check the box.

- ☐ The writer began with a sentence that clearly states the main idea of the summary.
- ☐ The sentences that follow give only the most important ideas.
- ☐ The writer put sentences in an order that makes sense.

C. How do you think the writer could make the summary better?

Possible responses: She could leave out her own opinions, or she could eliminate details. Words like *CHOMP* are unnecessary.

Writer's Companion ■ UNIT 2
Lesson 3 *More About Topic Sentence and Details*
54

See the rubric on page 207 for another way to evaluate the Student Model.

EXTENDING THE CONCEPT: FACT AND OPINION

Elaborate on the concept of fact and opinion by reminding children that a fact can be proven true or false but an opinion expresses a feeling, thought, or belief that cannot be proven true or false. Explain that a fact can be checked through resources such as experiments, experts, reference books and articles, and reliable online sites. Read these statements and have children tell which states a fact and which states an opinion: *The house is ten feet tall.* (Fact) *The house is the best one in town.* (Opinion)

Revise by Deleting

OBJECTIVES

- To revise a summary by deleting unimportant details
- To differentiate between important and unimportant ideas
- To revise an earlier piece of writing by utilizing only important details

Teach/Model Explain that when writers think about revising they think about what will make their writing clearer and stronger. Explain that there are many ways to revise. Tell children that, in this lesson, they will revise by deleting details and words that are unimportant in a summary.

Guided Practice Read aloud the introduction and example. Point out that the phrase "unimportant idea" means that the detail should be deleted from the summary on page 53. Have children turn back to the Student Model. Then ask why the example detail should be deleted. (It is an opinion and does not tell important information about the literature being summarized.) Guide children as they determine whether or not Item 1 in Part A is important to the summary on page 53.

Independent Writing Practice Have children complete the rest of Part A and Part B independently.

Name ______________________

Revising the Student Model

Revise by Deleting

One thing the writer could have done better is to delete details and words that are not important. Here is an example of how the Student Model can be improved.

Example It must have been so scary.

unimportant idea

A. The example above is an opinion and is not an important idea in the summary. Decide which sentences are important for the summary. Write *important idea* or *unimportant idea* on the line and explain why. Answers will vary. Possible responses are given.

1. Tortoises don't mind the spines of a cactus.

   unimportant idea–The sentence has nothing to do with the meeting.

2. She had to move out of the animal's way before it bumped right into her!

   important idea–The sentence shows how close they were.

3. There was just a camera between them.

   important idea–The sentence shows how close a nature photographer sometimes needs to get.

B. Now revise the summary you wrote on page 52. Delete unimportant details or those that do not support the topic. Do your writing on another sheet of paper.

Answers will vary. Students should restate the most important points of the paragraph they wrote.

WRITER'S STRATEGY: REVISION AND EDITING JOURNAL

Tell children they can make the revision and editing process more effective by keeping a personal revision and editing journal. Have them begin by dividing a page into two columns: *Strengths* and *Needs Improvement*. Then tell children to make comments in each column based on the revisions of their paragraphs on page 52. Explain that, as children keep this journal over time, they will be able to quickly spot and utilize strengths while working to improve other areas.

Review Organization

OBJECTIVES

- To review the terms *main idea, topic sentence, details, time-order words,* and *sequence*
- To identify main idea, topic sentence, details, time-order words, and sequence in literature
- To identify a detail that supports a topic sentence

Teach/Model Tell children that in this lesson they will review what they have learned about main idea, topic sentence, supporting details, time-order words, and sequence words. Read aloud the introduction. Read aloud the paragraph from *Sue, the Tyrannosaurus Rex*. Tell children to listen for the main idea, details, and time-order words.

Guided Practice For Item 1 in Part B, have children bracket an important detail about Sue. Remind them that the details tell about the main idea. Then guide children in completing Item 2 as they find and box a sentence that contains a time-order word. Make certain they circle the time-order word within the sentence.

Independent Writing Practice Discuss the directions and then have children complete Part C independently.

Use a Literature Model

LESSON 4: REVIEW ORGANIZATION

Name ______________________________

Writer's Craft in Literature

Review Organization

An effective paragraph has a **topic sentence** that tells the main idea. It has **details** that support the main idea. Effective paragraphs often use **time-order words** to show the **sequence** in which events happen.

A. **Read the paragraph about dinosaur fossils. The topic sentence is in the middle of the paragraph. It is underlined for you.**

Literature Model

[It took two years for a museum team to strip rock from the bones.] Team members also made models to replace some of the missing bones, using what they know from other dinosaur skeletons. Almost 90% of Sue's skeleton was found, which makes it the most complete *T. rex* skeleton today. Since the first discovery in 1900, twenty-two *T. rex* skeletons have been found. Only seven of those are more than half complete.

—from *Sue, the Tyrannosaurus Rex*
by Andrew Keown

B. **Review main idea, details, and time-order words.**

1. Put [ ] around an important detail about Sue.
2. Put a box around the sentence that contains a time-order word or phrase. Circle the time-order word or phrase.

C. **How were the missing bones replaced?**

Possible response: Team members made models from information they had from other dinosaur skeletons.

Writer's Companion ■ UNIT 2
Lesson 4 *Review Organization*
56

SHARING AND DISCUSSING

Have children share their responses to Part C. Ask if they got their information in a topic sentence or a detail sentence. Tell them to explain how the details support the main idea. Invite volunteers to add additional details.

Review Organization

OBJECTIVES

- To review how details in a paragraph support the main idea
- To review the concept that a topic sentence may appear anywhere in a paragraph
- To review time-order words in a paragraph

Teach/Model Read aloud the introduction. Remind children that a topic sentence tells the main idea in a paragraph. Use the graphic organizer to emphasize how details support the main idea by telling more about it. Read aloud the directions for Part A and the paragraph from *Sue, the Tyrannosaurus Rex*. Tell children to listen for the main idea and details. Explain that they should also think about sequence as they listen to the paragraph.

Guided Practice Guide children in identifying and underlining the topic sentence. Emphasize that a topic sentence does not always appear at the beginning of a paragraph. Then guide children in completing Part B.

Independent Writing Practice Have children complete Part C independently. Before answering, tell them to find the word *while* in the paragraph.

Name ______________________

A Closer Look at Writer's Craft

Review Organization

In this unit, you learned that an effective paragraph contains a topic sentence and details. Most topic sentences come at the beginning of a paragraph. Some, like the one below, are at the end. Others are even in the middle.

The **topic sentence** tells the main idea of the paragraph.

A **detail** supports the topic sentence and main idea. | A **detail** supports the topic sentence and main idea. | A **detail** supports the topic sentence and main idea.

A. Read this paragraph from *Sue, the Tyrannosaurus Rex*. Notice the topic sentence and the details that support it.

On a hot summer day in 1990, a woman named Sue Hendrickson was working with a team of fossil hunters near Faith, South Dakota. While some of the team went into a nearby town, Hendrickson stayed behind to look for fossils. What she found that day would be one of the greatest fossil discoveries ever made.

B. Use the paragraph to answer the questions below.

1. Underline the topic sentence of the paragraph.
2. Circle the main idea part of the topic sentence.
3. Box a detail sentence.

C. What happened *while* the fossil hunters went to a nearby town?

Possible response: Sue stayed behind to look for fossils.

____

____

WRITER'S GRAMMAR: COMMA AFTER INTRODUCTORY CLAUSE

Point out these introductory clauses in the paragraph:
On a hot summer day in 1990,
While some of the team went to a nearby town,
Explain that these groups of words tell *when*. They introduce the rest of the sentence and must be followed by a comma. This sets the introductory part of the sentence away from the rest of the sentence. The comma helps to make the sentence easier to understand.

Review Organization

OBJECTIVES

- To practice a prewriting technique to organize ideas
- To use time-order words to write steps in correct order
- To write supporting details that include time-order words

Teach/Model Tell children that it is important to plan before starting to write. Read aloud the text at the top of the page and have a volunteer read the content of the graphic organizer.

Guided Practice Read aloud the directions and Item 1 in Part A. Have children suggest time-order words to use in organizing a paragraph about Fiona cleaning her room. Then guide children in completing Item 2 by writing a main idea for the paragraph. Tell children to think about all the tasks as they determine the main idea.

Independent Writing Practice After reading aloud the directions, have children complete Part B independently. Tell them to think about two more details that tell more about the main idea. Remind them to include a time-order word in each sentence and to place sentences in an order that makes sense.

LESSON 4: REVIEW ORGANIZATION

Name ______________________

Practice with Writer's Craft

Review Organization

Before you write, you should organize your ideas. Here is how one third grader organized his ideas for a how-to essay.

Idea for the topic sentence:	Fiona has to clean her room.
Order	**Steps of the Task**
1	She picks up her clothing.
3	She puts her books on the shelf.
2	She dusts the furniture.
Conclusion:	Now her guests will see what a nice room she has.

A. Read the following topic sentence in bold type. Think of what you could do to turn it into an effective paragraph.

Today, Fiona will clean her room.

1. Put the steps of the task in the correct order. Use time-order words to show the sequence.

First, she picks up her clothing.

Next, she dusts the furniture.

Finally, she puts her books on the shelf.

2. What is the main idea of the paragraph?

Fiona cleans up her room.

B. Write two other details that could be added to the paragraph. Use time-order words to put them in the correct sequence.

Answers will vary, but should help describe the steps of the task in sequence.

____

____

Writer's Companion ■ UNIT 2
Lesson 4 *Review Organization* 58

Reaching All Learners

Below · On-Level · Advanced · ELL

BELOW LEVEL

Guide children in recognizing that the events in the example box are not in the correct order. Show children how the events can be placed in an order with the numbers next to them.

ADVANCED

Have children write a new paragraph. Tell them to include sequence words in a paragraph that tells what Fiona does after she cleans her room.

ENGLISH LANGUAGE LEARNERS

Pair English language learners with proficient English speakers. Have them work together to create a list of time-order words and then choose words from the list to complete Part B.

The Parts of a How-to Essay

OBJECTIVES

- To understand how a how-to essay is organized
- To analyze a Student Model

Teach/Model Read the introduction aloud. Explain that the writing model on this page is a good draft of a third grader's how-to essay. Read aloud the steps that explain how to organize this type of essay. Emphasize that it should include the necessary steps in time order. Then read the model aloud. Have children listen for the following: time-order words, the process being explained, and materials needed.

Guided Practice Point out the items below the Student Model. Read aloud the first item and guide children to identify and underline the first sentence in the essay. Next, guide children in boxing the list of materials (Item 2). Discuss that it is important for the writer to explain what will be needed in order to play the game.

Independent Writing Practice Have children complete Item 3 independently.

LESSON 4: REVIEW ORGANIZATION

Name ____________________

Focus on the Writing Form

The Parts of a How-to Essay

A **how-to essay** usually begins with a topic sentence about what is being explained. The writer then gives the steps, using time-order words that show the sequence to follow. Here is a draft of a how-to essay written by a third grader. As you read, think about how the student organized the essay. Then answer the questions.

Student Model

DRAFT

How to Play Island
by Marc

Here is how to play the game, Island. You need:

a CD player
lively music
9 sheets of colored paper
10 players

First, scatter the colored sheets of paper on the floor around the room. These are the "islands." Have someone start and stop the music. When the music starts, the players walk around the room. This is the fun part. When the music stops, each player jumps on an island. The player who is not on an island is "out." Remove an island. Then start the music again. Continue until there are two players and one island left. The last person standing on the island is the winner!

Introduce the essay with a topic sentence. It should describe the process you are explaining.

List the materials needed for the process.

Provide the steps in the order they should be done.

Use time order-words to show sequence.

Conclude by describing the result or outcome of the process.

1. Underline the sentence that tells what will be explained.
2. What materials are needed to play this game? Draw a box around them.
3. What time-order words show the sequence that should be followed?

First, When, Then, Continue until

Writer's Grammar
When you list materials, you can number them or put bullets (big dots) next to them. Go back and put a number or bullet next to each item on the material list in the Student Model.

59 **Writer's Companion** ■ UNIT 2
Lesson 4 *Review Organization*

WRITER'S GRAMMAR: TEXT FEATURES

Tell children that text features can make writing clearer and easier to understand. Explain that bullets—large, dark dots—are one type of text feature that can help readers find and understand a list. Point out the list of materials in the Student Model. Ask children how the Student Model was improved when they added bullets or numbers. Invite volunteers to tell when they think they might use bullets or numbers in their writing in the future.

Evaluate a How-to Essay

OBJECTIVES

- To review a student's how-to essay that received a score of 4 on a 4-point rubric
- To use teacher's comments to gain experience in evaluating a how-to essay
- To suggest improvements for a how-to essay

Teach/Model Explain that student writing in school and tests sometimes will be evaluated through a *rubric*, a chart that grades parts of the writing on a scale of points. Tell children that they will read a story that received a score of 4 on a 4-point rubric on this page.

Guided Practice Read aloud the directions for Part A. Then read the Student Model aloud. Have children listen (without looking at the essay) and think about what the writer did well and what needs improvement. Next, ask children to read the passage and the teacher comments silently. Guide children in discussing the teacher's comments and adding their own ideas regarding strengths and areas in the essay that need improvement. Ask children if they think the student writer did a good job of using interesting language.

Use a Student Model

LESSON 4: REVIEW ORGANIZATION

Name ______________________________

Evaluating the Student Model

Evaluate a How-to Essay

A. Two students were asked to write a how-to essay about something they know. This how-to essay got a score of 4. When using a 4-point rubric, a score of 4 means "excellent." Read the how-to essay and the teacher comments that go with it. Find out why this how-to essay is a success.

Student Model

How To Put Together a Jigsaw Puzzle
by Justin

Here's how I do a jigsaw puzzle. Before I start, I study the picture on the box. I spread out all the puzzle pieces on the floor, picture side up. Begin with the borders, because border pieces have a flat edge. After the borders, I look for any picture detail that stands out. Like if there's a haystack, I find all the pieces that could be a haystack. If there's a sun I look for all the pieces that could be a sun. Next, I go back to the border. If there are any weird shapes sticking out, I try to match those first. The background comes last, because it's the hardest to do. Sometimes it is just a blue sky. When that happens I take all the blue pieces and try to fit them together.

I work on three or four areas of the puzzle at the same time. When I have most of a puzzle done I look at the pieces and try to fit them in somewhere. Soon it starts to look like the picture on the box!

Good, you introduced the topic in the first sentence.

Nice work! You included good details that help explain the task.

Yes! You used time-order words like *next* to show the correct order of the steps.

You have described the result of the process and made it seem like fun too! Good job!

WRITER'S STRATEGY: SENTENCE VARIETY

Explain that children should use a variety of sentence lengths and types in their writing. Tell students this can make writing flow more easily, instead of reading like a boring, choppy piece. Have students point out two long sentences and two short sentences in the model. Then ask them to point out two sentences that have introductory clauses. Invite small groups to participate in a group activity to brainstorm a variety of sentence types and lengths they can use to write a short paragraph about how to do a favorite activity. Have groups share by writing their paragraphs and reading them to the class.

Evaluate a How-to Essay

OBJECTIVES

- To review a student's how-to essay that received a score of 2 on a 4-point rubric
- To use a teacher's comments to learn how to improve a how-to essay
- To suggest improvements for a how-to essay and use a rubric for evaluation

Teach/Model Tell children they will now see an example of a how-to essay that received a score of 2 (needs improvement) on a 4-point rubric. Read the Part B directions. Then tell children you will read the model. Explain that they should not read along with you but should listen closely and think about what the writer did well and what the writer might have done better.

Guided Practice Before they read the teacher comments, have children discuss strengths and areas that need improvement in the model. Then have students read the teacher comments. Invite a volunteer to summarize all the comments to tell why the essay needs improvement.

Independent Writing Practice Use the Extending the Concept feature below to help children use a rubric. Then have them complete the rubric independently.

Use a Student Model

LESSON 4: REVIEW ORGANIZATION

Name ______________________________

Evaluating the Student Model

B. This paragraph got a score of 2. Why did it get a low score?

Student Model

Peanut Butter Spaghetti
by Philip

I like to help my mom. Last night we made spaghetti. First, my mom made the pasta. She put it into a big bowl while it was still hot. But before she did that, I made the sauce. I put some tomato paste into a big bowl and added a little water and mixed it up for a long time. You have to keep on adding water a little at a time because peanut butter is hard to mix. Finally, I add spices. I like it spicy.

Can you introduce the subject in the first sentence?

Put the events in the correct order. This step should go first.

Nice use of time-order words.

We need to know the amounts of each ingredient.

This detail seems unnecessary. Can you write a better ending?

C. What score would you give the student's story? Put a number on each line.

	4	3	2	1
Introduction (Topic Sentence) ______	☐ The writer states the main idea in the topic sentence.	☐ The writer gives some idea of the topic.	☐ The writer only states the topic in the title.	☐ The writer does not state the topic.
Body (Details) ______	☐ The writer tells about many of the steps to follow.	☐ The writer tells about some of the steps to follow.	☐ The writer tells about few of the steps to follow.	☐ The writer does not tell about the steps to follow.
Time-order Words ______	☐ The writer uses many time-order words.	☐ The writer uses some time-order words.	☐ The writer uses few time-order words.	☐ The writer uses no time-order words.

EXTENDING THE CONCEPT: USING A RUBRIC TO EVALUATE

Direct children's attention to the rubric. Have a volunteer read each box in the *Introduction* row. Tell children to check the box that best tells how the student writer handled the introduction in the model. Then point out the number at the top of the column that contains the checked box. Tell children to write that number in the blank below the word *Introduction*. Have children repeat for *Body* and *Time-Order Words.*

Extended Writing/Test Prep

OBJECTIVES

- To apply the craft of paragraph organization to a longer piece of writing
- To select a topic for an extended piece of writing and begin the prewriting process
- To become familiar with types of writing prompts that appear on standardized assessment tests

Teach/Model Tell children that the next two pages will give them an opportunity to select a topic for a longer piece of writing. They will follow the steps of the writing process to complete it. Explain that children will choose from three possible writing activities. Discuss the choices listed in Part A and have each child select one.

Guided Practice Review the writing forms covered in the theme. Then have children reread the Student Model tips in the lesson where they initially learned about the chosen writing form:

- Directions: Lesson 1, Page 41
- A Paragraph of Information: Lesson 2, Page 47
- A Summary: Lesson 3, Page 53
- A How-to Essay: Lesson 4, Page 59

Have children begin the prewriting process as you guide them in completing Part B.

LESSON 4: REVIEW ORGANIZATION

Name ______________________________

Extended Writing/Test Prep

Extended Writing/Test Prep

On the last two pages of this lesson, you will use what you have learned about effective paragraphs to write a longer written work.

A. Read the three choices below. Put a star by the writing activity you would like to do.

1. Respond to a Writing Prompt

   Writing Situation: You have to explain to a friend how to make your favorite sandwich.

   Directions for Writing: Think about your favorite kind of sandwich. Now write a how-to essay that will explain how to make it. Use what you have learned about topic sentences, details, and time-order words.

2. Choose one of the pieces of writing you started in this unit:
   - directions (page 40)
   - a paragraph of information (page 46)
   - a summary (page 52)

   Expand your beginning into a complete piece of writing. Use what you have learned in your writing.

3. Choose a topic you would like to write about. You may explain how to do an activity you know.

B. Use the space below and on the next page to plan your writing.

TOPIC: ______________________________

____

WRITING FORM: ______________________________

HOW WILL I ORGANIZE MY WRITING: ______________________________

____

____

____

____

Writer's Companion • UNIT 2
Lesson 4 *Review Organization*
62

EXTENDING THE CONCEPT: CHOOSING A TOPIC

Tell children they can most effectively choose a topic by asking themselves questions such as these: What do I know about the topic? How much would I enjoy writing about the topic? Do I have enough to say about the topic to write a whole composition about it? Based on what I think and know, do I have enough time to organize a composition on this topic? Suggest that children circle keywords in the directions to help them answer these questions.

Extended Writing/Test Prep

OBJECTIVES

- To use a graphic organizer in preparing to create a longer piece of writing
- To use the steps of the writing process to complete a longer piece of writing independently

LESSON 4: REVIEW ORGANIZATION

Name ____________________

Extended Writing/Test Prep

C. In the space below, draw a graphic organizer that will help you plan your writing. Fill in the graphic organizer. Use the lines below to write any notes you have.

Notes

____

____

____

____

____

____

____

____

D. Do your writing on another sheet of paper.

63

Writer's Companion ■ UNIT 2
Lesson 4 *Review Organization*

Teach/Model Read aloud the directions for Part C and explain to children that they will choose a graphic organizer to help them plan their writing. Provide each child with a photocopy of the appropriate graphic organizer from the back pages of this guide. Model how to complete one of the graphic organizers.

Guided Practice Guide children in completing their chosen graphic organizers. Remind them that a graphic organizer is a planning tool and need not include every detail. Explain that children can add additional details as they write.

Independent Writing Practice: Draft, Revise, Publish Tell students to use the graphic organizers as they write on a separate sheet of paper. After the first draft is complete, tell students to revise while considering organization and word choice. Organize a display of student writing.

SHARING AND DISCUSSING

Have children discuss how they chose the graphic organizer they decided to use. Ask them to explain why and how this specific graphic organizer helped them prepare to write. Then have children point out specific portions of their writing and explain which part of the graphic organizer they used to write each one.

Answering Multiple-Choice Questions

OBJECTIVES

- To understand that some writing tests contain multiple-choice questions
- To strategically utilize knowledge of compound sentences before making an answer selection

Teach/Model Tell children that some writing tests include multiple-choice questions. Explain that this lesson will help children develop the ability to read and answer this type question well. Read aloud the directions for Part A. Then read aloud the test tip. Model answering Question 1 by writing the item on the board. Circle the boldfaced word in the question. Circle the commas and end punctuation in the answer choices as you point out the conjunction *but* in each one. Identify the punctuation in each answer choice as you model how to determine the correct answer.

Guided Practice Have a volunteer read aloud Question 2, making certain children note the boldfaced word in this question. Then tell children to silently read the answer choices and mark the correct answer.

Independent Writing Practice Have children complete Questions 3–4 independently.

Use a Literature Model

LESSON 5: WRITING TEST PRACTICE

Name ______________________________

Writing Test Practice

Answering Multiple-Choice Questions

Some writing tests have questions with answer choices. This lesson will help you practice this kind of test.

A. Some test questions may ask you about sentences. Read the test tip. Then read each question. Fill in the circle next to your answer.

Read the questions below. Answer questions.

Test Tip: Compound sentences are joined with a comma and a conjunction such as *and* or *but*.

1. In which sentence below is all **punctuation** correct?
   - (A) I like broccoli but I don't like beets.
   - (B) I like broccoli, but I don't like beets. (filled)
   - (C) I like broccoli, but I don't like beets?
2. In which sentence below is all **capitalization** correct?
   - (F) I have a cat named Buster. (filled)
   - (G) I have a cat named buster.
   - (H) I have a Cat named Buster.
3. In which sentence below is all **punctuation** correct?
   - (A) We saw geese and chickens. (filled)
   - (B) We saw geese, and chickens.
   - (C) We saw geese and chickens?
4. In which sentence below is all **capitalization** correct?
   - (F) susan and her brother are very much alike.
   - (G) Susan and her brother are very much alike. (filled)
   - (H) Susan and her Brother are very much alike.

Writer's Companion ▪ UNIT 2
Lesson 5 *Writing Test Practice*
64

SHARING AND DISCUSSING

Have pairs discuss their answers to Questions 3 and 4. Tell them to explain how they worked through the answer choices and why they made their answer selections.

Answering Multiple-Choice Questions (cont.)

OBJECTIVES

- To understand that some multiple-choice questions require that students first read a passage
- To become familiar with reading numbered sentences in a passage
- To learn the strategy of answering questions on an Answer Sheet

Teach/Model Tell children that some tests ask them to read a passage and then answer questions about it on an Answer Sheet. Read aloud the directions and passage in Part B. Explain that the sentences are numbered because the numbers are used in the questions that follow the passage. Read aloud the first question and model returning to the passage to find Sentence 1. Then model selecting the correct answer and correctly using the Answer Sheet.

Guided Practice Guide children as they complete Question 2.

Independent Writing Practice Have children complete Question 3 independently. Then read aloud and discuss the test tip. Remind children that a compound sentence can be broken into two sentences and have children complete Question 4 independently.

LESSON 5: WRITING TEST PRACTICE

Name ______________________

Writing Test Practice

B. **Some tests may ask you to read a passage and answer questions about it. The sentences in the passage may be numbered. Read the test tip. Then answer the questions.**

Directions: Read the passage below. Then read questions 1–4 and mark the correct answers on your Answer Sheet.

(1) There are twenty child in my class. (2) My best friend's name is Marianna. (3) Sometimes we do homework together. (4) She has two parrot. (5) One parrot's name is Polly, and there is another parrot whose name is Cracker.

Test Tip: A compound sentence can be broken into two complete sentences.

1. What change, if any, should be made in sentence 1?
   - A Change ***child*** to **childs**
   - B Change ***child*** to **children**
   - C Change ***class*** to **classes**
   - D Make no change
2. Which, if any, is the correct way to revise sentence 2?
   - F My best friend's name is marianna.
   - G My best friends name is Marianna.
   - H My best friend's names is Marianna.
   - J No change is needed.
3. Which, if any, is the correct way to revise sentence 4?
   - A She has two parrots.
   - B She has two parrotes.
   - C She has two parrottes.
   - D No change is needed.
4. Which would make sentence 5 the **BEST** compound sentence?
   - F One parrot's name is Polly, or the other is Cracker.
   - G One parrot's name is Polly and the other is Cracker.
   - H One parrot's name is Polly, and the other parrot's name is Cracker.
   - J One parrot's name is Polly and the other parrot's name is Cracker.

Answer all test questions on this Answer Sheet.

1.	Ⓐ	**B** (filled)	Ⓒ	Ⓓ	3.	**A** (filled)	Ⓑ	Ⓒ	Ⓓ
2.	Ⓕ	Ⓖ	Ⓗ	**J** (filled)	4.	**F** (filled)	Ⓖ	Ⓗ	Ⓙ

65 Writer's Companion • UNIT 2 Lesson 5 *Writing Test Practice*

USING ACADEMIC LANGUAGE

Children will usually encounter academic language on tests. This is not the language they use in conversation, but it is a very specific type of language they must work with to understand directions, questions, and some answer options. Prepare children by helping them recognize and understand important words and phrases in directions. Explain that the phrases *passage below* and *following is a paragraph* mean children are getting ready to read something the questions will ask them about.

Answering Multiple-Choice Questions (cont.)

OBJECTIVES

- To practice answering multiple-choice questions based on a reading passage
- To practice identifying errors in a passage and using an Answer Sheet
- To understand that each type of question following a passage can be different

Teach/Model Explain that each question following a passage can ask a different kind of question. Questions on this sample test ask which word is not written correctly and the correct way to write a sentence. Have students read the passage silently. Then model thinking through answer choices and answering Item 1 on the Answer Sheet.

Guided Practice Read aloud the test tip and model answering Item 3 by reading the answer choices. Make certain children understand that *Leave as is* means "make no changes."

Independent Writing Practice Have students answer Item 2 independently on the Answer Sheet.

LESSON 5: WRITING TEST PRACTICE

Name ______________________

Writing Test Practice

C. Some test questions may ask you the best way to revise or correct sentences in a paragraph. Read the test tip. Then answer the questions on the Answer Sheet.

The following is a paragraph that may contain errors.

My Favorite Day

(1) My favorite day of the week is saturday. (2) That is when I go to the market with my dad. (3) we go there every saturday morning. (4) We get there early. (5) The farmer's are already there. (6) I would like to be a farmer someday. (7) It must be hard work!

1. Which underlined word is *not* written correctly?
   - A My favorite
   - B saturday
   - C my dad
   - D We get there early.
2. Which is the correct way to capitalize sentence 3?
   - F We go there every saturday morning.
   - G We go there every Saturday Morning.
   - H we go there every Saturday morning.
   - J We go there every Saturday morning.
3. How should the underlined words in sentence 5 be correctly written?
   - A The farmers
   - B The farmers's
   - C The farmers'
   - D Leave as is.

Test Tip: To answer question 3, remember that a word that shows ownership contains *'s* or *s'*.

Answer all test questions on this Answer Sheet.

1. Ⓐ ● Ⓒ Ⓓ
2. Ⓕ Ⓖ Ⓗ ●
3. ● Ⓑ Ⓒ Ⓓ

Writer's Companion ▪ UNIT 2
Lesson 5 *Writing Test Practice* 66

Reaching All Learners

BELOW LEVEL

Suggest that children take enough time, within the constraints of the test, to answer. Tell them it is fine if others finish before they do. This can help allay the anxiety that might distract them from succeeding.

ADVANCED

Invite children to write a new short passage and create items that address the same skills addressed in the questions on this page. As they build errors into a passage and create answer choices, children will naturally grow to better understand the testing process.

ENGLISH LANGUAGE LEARNERS

ELL children may grow frustrated quickly when reading a passage. Suggest they relax as they read the passage and then read the questions to see which parts of the passage are tested. Then they can focus on the tested portions, as well as important words such as *not* and *capitalize*.

Answering Multiple-Choice Questions (cont.)

OBJECTIVES

- To understand that some passages contain blanks to be filled with answer choices
- To practice filling in choices next to the correct answer

Teach/Model Explain that some passages contain blanks and that each blank will be filled by an answer choice. Make certain children understand that on some tests they will not write in the blank. Instead, they will fill in bubbles next to the correct answer. Have children read the passage silently. Then model answering Item 1 by reading the answer choices, returning to the passage for context, and determining which answer choice is correctly punctuated. Emphasize how to fill in the correct answer bubble.

Guided Practice Read aloud the test tip and guide children as they complete Item 2, emphasizing that the sentence does not need to show possession, just a plural. Model answering Question 1 by reminding students they must again return to the passage for context.

Independent Writing Practice Have children complete Items 3 and 4 independently.

LESSON 5: WRITING TEST PRACTICE

Name ______________________

Writing Test Practice

D. Some test questions may ask you about parts of sentences. Read the test tip. Then answer the questions.

Read the paragraph. Read each question and mark the circle next to the correct answer.

Yesterday our class went to the zoo in __(1)__. First, we saw the seals. We saw them eat their lunch. Two __(2)__ fed them. The seals did tricks. After lunch they took a nap, __(3)__ we went to see the birds. One __(4)__ feathers were all white.

1. Which answer should go in blank (1)?
   - Ⓐ jacksonville
   - Ⓑ Jacksonville
   - Ⓒ zoo
2. Which answer should go in blank (2)?
   - Ⓕ woman
   - Ⓖ womans
   - Ⓗ women
3. Which answer should go in blank (3)?
   - Ⓐ or
   - Ⓑ so
   - Ⓒ unless
4. Which answer should go in blank (4)?
   - Ⓕ birds
   - Ⓖ bird's
   - Ⓗ birds'

Test Tip: When more than one person or thing owns something else, it is shown with an *s'*. For instance, *pigs'* in *the pigs' tails*, tells us that there is more than one pig and each pig has a tail.

67 Writer's Companion • UNIT 2 Lesson 5 *Writing Test Practice*

ASSESSING STUDENT RESPONSES

If a child answers test items incorrectly, consider possible causes for the incorrect responses. One cause might be a failure to return to the passage before answering. For example, in Item 3, all answer choices are words that could be correct in a specific context. If a child is missing this type of question frequently, you might ask the child to walk you through their answering process. Then you will be able to focus on the type of error and provide examples to point out the importance of returning to a passage to see exactly how a tested word is used.

Look at Supporting an Opinion

OBJECTIVES

- To understand the term *opinion*
- To identify opinions in literature

Teach/Model Read aloud the introduction and tell students that an opinion expresses how you feel or what you think about something. Explain that opinions are personal, and they depend on whom is speaking or writing. Now read aloud the directions for Part A and the passage from *The Stories Julian Tells.* Ask students to listen and pay attention to Julian's feelings about friends who are girls and about Huey's friend.

Guided Practice Guide students in completing Part B. Ask students how Julian feels about friends who are girls. Identify Julian's opinion from the first paragraph and have students underline it. (*I didn't want a girl for a friend.*) Then have students identify Julian's second opinion from the passage (Part B, Item 2). (*I think he is too young to play with.*)

Independent Writing Practice After reading aloud the directions, have students complete Part C independently. When they have finished, have them share and discuss the words they have identified.

Use a Literature Model

LESSON 1: SUPPORTING AN OPINION

Name ______________________

Writer's Craft in Literature

Look at Supporting an Opinion

Sometimes when you write, you will express an **opinion**. An opinion tells how you feel or what you think about something.

A. Read the following passage from *The Stories Julian Tells.* Notice how Julian feels about friends who are girls and about Huey's friend.

Literature Model

If you have a girl for a friend, people find out and tease you. That's why I didn't want a girl for a friend — not until this summer, when I met Gloria.

It happened one afternoon when I was walking down the street by myself. My mother was visiting a friend of hers, and Huey was visiting a friend of his. Huey's friend is five and so I think he is too young to play with.

—from *The Stories Julian Tells*
by Ann Cameron

B. Identify details from the passage that show opinions.

1. Underline Julian's opinion in the first paragraph.
2. Circle Julian's opinion in the second paragraph.

C. Write the words that tell why Julian feels the way he does.

Possible response: People find out and tease you; Huey's friend is five

Writer's Companion ▪ UNIT 3
Lesson 1 *Supporting an Opinion* 68

SHARING AND DISCUSSING

Have students compare and discuss Julian's opinions from the passage. What do these opinions say about Julian? Do you agree or disagree with Julian's opinions?

Explore Supporting an Opinion

OBJECTIVES

- To identify an opinion and its supporting reasons and details
- To use reasons and details in supporting an opinion

Teach/Model Read aloud and explain the diagram at the top of the page in order to help students understand that an opinion is supported by reasons and details.

Guided Practice Tell students that supporting reasons and details help prove and support an opinion. Read aloud the directions for Part A and complete the example on the board. Tell students to ask themselves if the sentence is an opinion. Does it tell how one is feeling? Or, what one thinks about something? Then continue guiding students until Items 1–3 are completed.

Independent Writing Practice Read aloud the directions for Part B and the paragraph. Then have students follow along on their own page. Have students complete Part B independently. For Part C, see *Extending the Concept*.

Name ______________________

A Closer Look at Writer's Craft

Explore Supporting an Opinion

You can support your opinions by giving **reasons** or **details**.

A **reason** tells why you feel that way.

An **opinion** tells how you feel or what you think about something.

Details give more information about your opinion.

A. Read each sentence. Circle whether each is an opinion or a supporting detail.

Example Winter is the best time of year.
DETAIL (OPINION)

1. The swimming pool is five miles away.
(DETAIL) OPINION

2. Going to the movies is really fun.
DETAIL (OPINION)

3. The woods are scary at night.
DETAIL (OPINION)

B. Read the paragraph. Think about the opinion. Underline the reasons that support the opinion.

People should keep their dogs on leashes in the park. There is a leash law, but people don't follow it. A dog off the leash can be a danger. It can bite someone or knock a child down. It can also get in fights with other dogs.

C. Think of another reason that supports the opinion. Write it on the lines.

Possible response: A dog off the leash can get lost.

____

EXTENDING THE CONCEPT: REASONS, REASONS, AND MORE REASONS

Before students complete Part C, have them think of several additional reasons why "people should keep their dogs on leashes in the park." Pair students and give them a few minutes to come up with a couple of new reasons. Then ask students to share their reasons and write them on the board. After discussing these possibilities, have students complete Part C independently.

Use Supporting an Opinion

OBJECTIVES

- To prepare to write a persuasive paragraph by writing an opinion and supporting reasons
- To practice using reasons to support an opinion

Teach/Model Tell students that before they write a persuasive paragraph, they should plan what they want to say. Read aloud the introduction at the top of the page. Then have students read the opinion and reasons in the example and identify whether or not each reason supports the given opinion.

Guided Practice After reading aloud the directions for Part A, ask students what kind of animal they think makes the best pet. Guide them in writing their opinions on the line in Part A and also filling in their charts. Remind students that they can turn back to page 69 if they need to review the meaning of the terms *opinion* and *reason*.

Independent Writing Practice After reading aloud the directions, have students complete Part B independently.

LESSON 1: SUPPORTING AN OPINION

Name ____________________

Practice with Writer's Craft

Use Supporting an Opinion

To persuade means to get someone to do something or to think in a certain way. Here is how one third grader planned a paragraph that persuades.

Example My Opinion: My dog is the best pet in the world.

Reason 1	Reason 2	Reason 3
He is always ready to play.	He listens when I tell him my problems.	He guards the house.

A. What kind of animal do you think makes the best pet? Write your opinion on the line. Then complete the chart with reasons that persuade.

My Opinion: ____________________

Reason 1	Reason 2	Reason 3

B. Use information from your chart to write a paragraph that states and supports your opinion. Do your writing on another sheet of paper.

Answers will vary, but should include an opinion and three reasons that support the opinion.

Writer's Companion ■ UNIT 3
Lesson 1 *Supporting an Opinion* 70

Reaching All Learners

BELOW LEVEL

Help students list reasons by prompting them with specific, guiding questions, such as "Why do you think your selected animal makes the best pet?" Then encourage students to complete their charts by following the given example.

ADVANCED

Encourage students writing above grade level to use a thesaurus to help them select more specific, expressive words to describe why they think their selected animal makes the best pet.

ENGLISH LANGUAGE LEARNERS

Pair less fluent English speakers and more fluent English speakers. Have students discuss which animals make the best pets. Have them help each other identify the appropriate English words and phrases to include in their charts.

The Parts of a Persuasive Paragraph

OBJECTIVES

- To understand how a persuasive paragraph is organized
- To analyze a Student Model

Teach/Model Read aloud the introduction and tell students that the writing model on this page is a first draft of a third grader's persuasive paragraph. Read aloud the steps for how to organize a persuasive paragraph. Then read aloud the model. As students listen, have them pay careful attention to how the reasons and details are organized.

Guided Practice Point out the questions below the Student Model. Read aloud the first question. Guide students in identifying and underlining the opinion. Remind students that an opinion expresses how you feel or what you think about something.

Independent Writing Practice Have students complete Items 2–3 independently. Then discuss their responses as a class.

Use a Student Model

LESSON 1: SUPPORTING AN OPINION

Name ____________________

Focus on the Writing Form

The Parts of a Persuasive Paragraph

A **persuasive paragraph** uses reasons and details to support a strong **opinion.** Below is an example of a persuasive paragraph written by a third grader. As you read, think about how the student supported her opinion. Then answer the questions.

Student Model

DRAFT

My Favorite Pet
by Marci

Gully is my dog, and he is the best pet in the world. He is great at listening. All day and all night, he guards the house. No matter what is bothering me, he makes me feel better. I always feel safe with him. He is always ready to play, so I never get bored. He is also very smart, much smarter than my friends' dogs. He'll play with a ball, a toy, or even a sock. He obeys many different commands.

Introduce the opinion in a clear topic sentence.

Organize the reasons and details in an order that makes sense.

Develop the argument in sentences that present your reasons.

Conclude by restating the opinion or calling for action.

1. Which sentence states an opinion? Underline it.
2. Which sentences give reasons? Put boxes around them.
3. Which sentences seem to be out of order? Circle them. Then draw an arrow to the place where you think each sentence belongs. Rearranging may vary.

Writer's Grammar
In a plural possessive noun, the apostrophe comes last. For example: *Students' books* or *dogs' bones*. Find plural possessive nouns in the Student Model.

71 Writer's Companion • UNIT 3
Lesson 1 *Supporting an Opinion*

WRITER'S GRAMMAR: AVOIDING REPETITION

Plural means "more than one." Point out that, in English, most nouns are made plural by adding *–s* or *–es* at the end. For example, the plural of tree is *trees* and the plural of brush is *brushes*. Then remind students that an apostrophe is used to show possession or ownership. Explain that, for a plural noun, the apostrophe is put after the final *–s*:

The students washed the *teachers'* cars.
The *birds'* calls woke us up.

Evaluate a Persuasive Paragraph

OBJECTIVES

- To understand what to look for when evaluating a persuasive paragraph
- To evaluate a persuasive paragraph
- To suggest improvements for a persuasive paragraph

Teach/Model Discuss with students why it is important to evaluate writing. Explain that writing is rarely finished after the first draft, so students should expect to evaluate first drafts and sometimes other drafts in order to make changes that will improve their writing. Using the points at the top of the page as a guide, discuss what to look for when evaluating a persuasive paragraph.

Guided Practice Read aloud the directions for Part A. Guide students in writing responses and referring to the Student Model for information. Tell students that using a checklist is helpful in evaluating writing because it allows them to focus on a few important elements. Point out the checklist in Part B and explain that it contains many of the writing skills practiced in this lesson.

Independent Writing Practice After reading aloud both sets of directions, have students complete Parts B and C independently.

Name ______________________________

Evaluating the Student Model

Evaluate a Persuasive Paragraph

When you evaluate a persuasive paragraph, ask yourself these questions:

- Does the writer express a strong opinion?
  (Look for signal words such as *think, believe, feel, best,* and *worst.*)
- Does the writer place reasons in a logical order?
  (Look for an order that makes sense.)

A. Reread the Student Model on page 71. Then answer these questions.

1. How does the writer feel about her dog?
   She thinks her dog is the best pet in the world.
2. Name two reasons why the writer feels that way.
   The dog is great at listening; he guards the house; he is always ready to play; he is very smart

B. Now evaluate the Student Model. Put a check in the box next to each thing the writer has done well. If you do not think the writer did a good job with something, do not check the box.

- ☐ The writer introduced the opinion in a strong topic sentence.
- ☐ The reasons and details are in a logical order.
- ☐ The writer included at least three reasons.
- ☐ The writer concluded by restating the opinion or calling for action.

C. How do you think the writer could make the persuasive paragraph better? Write your ideas below.

Possible response: The writer should put reasons in a more logical order. She could also conclude by restating the opinion.

See the rubric on page 207 for another way to evaluate the Student Model.

EXTENDING THE CONCEPT: WORD CHOICE

Point out the words *great* and *best,* which appear in the beginning of the Student Model. Explain that these words are general, unclear, and not very helpful to the reader. Write the following sentences on the board and have students work in pairs to brainstorm three clear, specific alternatives for each underlined word. Invite them to share their answers with the class.

- The meal was great.
- My friend is the best.

Revise by Rearranging Sentences

OBJECTIVES

- To revise a persuasive paragraph by rearranging sentences

Teach/Model Tell students that revising writing means changing it to make it better. Explain that there are many types of changes writers can make. They can change words and sentences or even change the organization and ideas. In this activity, students will revise the Student Model and their own writing by rearranging sentences.

Guided Practice Read aloud the introduction and the example. Discuss with students how moving *He is great at listening* makes the ideas in the model flow in a more logical order. Then read aloud the directions for Part A and guide students as you complete Item 1 together. After you have finished, complete Item 2 as a class.

Independent Writing Practice After reading aloud the directions, have students complete Part B independently. Invite volunteers to read their revisions aloud.

LESSON 1: SUPPORTING AN OPINION

Name ______________________

Revising the Student Model

Revise by Rearranging Sentences

When you revise a persuasive paragraph, you want to put your reasons in a logical order. Sometimes, this means that you need to move, or rearrange, your sentences. Here is an example of how a part of the Student Model can be improved.

Example The sentence *He is great at listening* should be moved close to *He obeys many different commands.*

A. **Revise these groups of sentences from the Student Model. Put them in an order that makes more sense to you.** Possible responses are given.

1. All day and all night, he guards the house. No matter what is bothering me, he makes me feel better. I always feel safe with him.

   All day and all night, he guards the house. I always feel safe with him. No matter what is bothering me, he makes me feel better.

2. He is always ready to play, so I never get bored. He is also very smart, much smarter than my friends' dogs. He'll play with a ball, a toy, or even a sock.

   He is always ready to play, so I never get bored. He'll play with a ball, a toy, or even a sock. He is also very smart, much smarter than my friends' dogs.

B. **Revise the paragraph you wrote on page 70. Focus on writing a clear opinion with reasons and details that support your opinion. Rearrange your sentences so they make the most sense. Do your writing on another sheet of paper.**
Answers will vary. Students should revise sentences so they are in a logical order.

73 Writer's Companion ■ UNIT 3
Lesson 1 *Supporting an Opinion*

WRITER'S STRATEGY: FOCUS ON A FEW CHANGES AT A TIME

Explain to students that, before they revise their work, they should determine two or three areas of focus. For example, in Part B, the focus was rearranging sentences into the best logical order. Sometimes a given checklist that includes areas of focus can be useful. Other times, students should take a few moments to develop their own checklist in order to guide themselves while revising their own work.

Look at Staying on Topic

OBJECTIVES

- To understand the term *stay on topic*
- To identify if sentences are on or off the topic in literature

Teach/Model After reading aloud the introduction, tell students that it is important for writers to *stay on topic*. This means that all the sentences in a paragraph should be about the writer's chosen topic. Read aloud the directions for Part A and the passage from *The Talent Show*. As students listen, ask them to pay attention to how all the sentences tell about the topic.

Guided Practice Read aloud the directions for Part B. Then ask a volunteer to read aloud Item 1. Guide students in identifying the topic sentence as the first sentence in the paragraph. Then go through the paragraph, sentence by sentence, with students, asking: "Is this sentence about the topic?" Have students underline the sentences that deal with the topic. For Item 3, have students put a box around the sentence that is off topic.

Independent Writing Practice After reading aloud the directions, have students complete Part C independently.

Use a Literature Model

LESSON 2: STAYING ON TOPIC

Name ______________________

Writer's Craft in Literature

Look at Staying on Topic

When you write, it is important to **stay on topic**. To do this, you must make sure all the sentences in a passage are about the topic of the passage.

A. Read the following passage. Notice how the writer's sentences tell about the topic.

Literature Model

On Friday our class put on the best talent show in the whole world. For his talent, Boomer Fenton showed his birthmark in the shape of a dog's face. Kelsey played "Twinkle, Twinkle, Little Star" on her violin. Leo tried to get his dog to roll over, but the dog ran under Ms. Babbitt's chair and wouldn't come out for the rest of the show. Carol Ann and Wendy did the bee poem. Carol Ann's crown fell off right in the middle of it.

—from *The Talent Show* by Susan Wojciechowski

B. Decide if the sentences are on or off the topic.

1. Circle the sentence that tells the topic.
2. Underline the sentences that are about the topic.
3. Choose which one of the sentences below is off the topic of the passage. Put a box around it.

Linda sang a song in a very high voice.

Doug went on a trip to Oregon with his father.

C. Reread the passage above. What is the author's opinion about the talent show.

Possible response: She thought it was the best talent show in the world.

____

Writer's Companion ■ UNIT 3
Lesson 2 *Staying on Topic* 74

SHARING AND DISCUSSING

Have students discuss how they were able to tell which sentences from the Literature Model stayed on topic. Invite volunteers to share their thoughts with the class.

Explore Staying on Topic

OBJECTIVES

- To identify if sentences are *on topic*
- To write a sentence that stays on topic

Teach/Model Using the diagram at the top of the page, explain the meaning of *staying on topic*. Review the definition of the terms *topic sentence* and *supporting details* and then discuss how supporting details should relate to the topic.

Guided Practice Read aloud the directions for Part A. Tell students that this activity will help them practice recognizing when sentences are on or off topic. Read aloud the example. Talk about the topic (*baking a pie*) and ask students to identify the sentence that is *not* about baking a pie. Then complete Items 1–2 as a class.

Independent Writing Practice After reading aloud both sets of directions, have students complete Parts B and C independently. When students are done, review their responses as a class.

LESSON 2: STAYING ON TOPIC

Name ______________________________

A Closer Look at Writer's Craft

Explore Staying on Topic

To stay on topic, your sentences all need to support your topic.

Staying on Topic

The **topic sentence** states the main idea.

Detail sentences suppport the topic sentence.

A. **Read the sentences below. Think about which ones are on topic and which ones are not. Cross off the sentence in each group that is not on topic.**

Example Tomás decided to bake a pie. He got out the sugar, the butter, and the flour. ~~He watched a show on television.~~

1. Rob really wanted to get on the baseball team. ~~He was taller than the other boys in his class.~~ He practiced batting and catching until he was tired.
2. My brother is the kindest person I know. He always helps around the house. He helps anyone in trouble. ~~He gets very good grades.~~

B. **Read the paragraph. Think about the topic. Cross out the sentence that is not on topic.**

The storm raced across the prairie. The sky got darker and darker. ~~The fields were turning green.~~ The wind began to blow. Suddenly, the clouds opened, and rain poured down.

C. **Write another detail sentence that could tell about the storm.**

Possible response: It rained so hard that the river overflowed its banks.

75

Writer's Companion • UNIT 3
Lesson 2 *Staying on Topic*

Reaching All Learners

BELOW LEVEL

For Part B, have students work with a partner or aide. Have students verbally describe details telling about *the storm.* Then have the partner or aide help students write a detail sentence about the storm.

ADVANCED

Encourage students writing above grade level to write three or four detail sentences about the storm.

ENGLISH LANGUAGE LEARNERS

For Parts B and C, pair less fluent English speakers with more fluent English speakers. Have students discuss possible details about *the storm.* Then have them identify appropriate English words and phrases in order to write a detail sentence about the storm.

Use Staying on Topic

OBJECTIVES

- To prepare to write a speech by writing a topic sentence that states an opinion
- To practice writing detail sentences that stay on topic

Teach/Model Tell students that, before writing a speech, they should spend some time planning what they want to say. Read aloud the introduction at the top of the page. Guide students through the example. After reading aloud the topic sentence, ask students to read aloud the details.

Guided Practice Read aloud the directions for Part A. Have students write a topic sentence about something they enjoy doing and want to persuade others to do. Guide them in filling out their charts with details that stay on topic.

Independent Writing Practice After reading aloud the directions, have students complete Part B independently.

LESSON 2: STAYING ON TOPIC

Name ____________________

Practice with Writer's Craft

Use Staying on Topic

Speeches are often written to persuade people to believe or to do something. When you plan a speech that persuades, start with a topic sentence that states your opinion. Then list some details about your topic. Here is how one student planned a speech about doing yoga.

Example

Topic Sentence	Yoga is a great activity that everyone should try.
Details	Yoga is relaxing Helps you get stronger Stretches muscles Good for every age group

A. Think about something you like to do and want to persuade others to do. Write a topic sentence. Write details that tell about your topic.

Topic Sentence	
Details	

B. Use information from your chart to write a speech that persuades. Remember to stay on topic. Do your writing on another sheet of paper.
Answers will vary, but should include detail sentences that tell about the topic.

Writer's Companion ▪ UNIT 3
Lesson 2 *Staying on Topic* 76

EXTENDING THE CONCEPT: OFF-TOPIC DETAILS

After completing Part A and before writing a persuasive speech in Part B, have students write a second list of details. Tell students that these details should be *off topic.* Ask them to write down four or five off-topic details. Then have students compare their lists of details. Tell students that this exercise is meant for them to continue practicing identifying if details are on or off topic.

The Parts of a Speech

OBJECTIVES

- To understand how a speech is organized
- To analyze a Student Model

Teach/Model After reading aloud the introduction, tell students that the writing model on this page is a speech written by a third grader. Explain that a good speech tries to convince the audience to believe something or to do something. Read aloud the comments about how to organize a speech. Then read aloud *A Perfect Activity.* As students listen, ask them to pay careful attention to any sentence that is off topic.

Guided Practice Point out the items following the Student Model. Read aloud the first item. Guide students in identifying and underlining the topic sentence. Then answer Item 2 as a class.

Independent Writing Practice Have students complete Items 3–4 independently. Then discuss their responses as a class.

Use a Student Model

LESSON 2: STAYING ON TOPIC

Name ______________________

Focus on the Writing Form

The Parts of a Speech

A good **speech** tries to persuade the audience to believe something or to do something. Below is a draft written by a third grader. As you read, think about how the student organized it. Then answer the questions.

Student Model

DRAFT

A Perfect Activity
by Maria

Yoga is a great activity that everyone should try. Part of each yoga class is spent just thinking and breathing. It can really relax you. You might be worried about something at the start of a yoga class. By the end, you will feel calm. Yoga can help you get stronger. ~~Other sports help you get stronger, too.~~ It is a good way to stretch your whole body. It works every muscle you have. Even your face and your feet will feel stretched. People of every age can take yoga. There are classes for seniors, and there are classes for kids.

Get the audience's attention with a strong statement of your opinion.

State reasons that support your opinion.

Add details that will appeal to your audience. Be sure the details are on topic. Put them in a logical order.

Conclude by restating your opinion and calling your listener to action.

1. Which sentence states the topic? Underline it.
2. Find one sentence that tells more about the topic. Put a box around it. Possible response
3. Circle one sentence that seems to be out of order. Then draw an arrow to the place where you think this sentence belongs.
4. Which sentence is not on topic? Cross it out.

Writer's Grammar
An action verb describes an action (*Crystal **runs** fast.*) and a linking verb connects the subject to information about the subject (*The kite **is** red.*). The Student Model contains both kinds of verbs. Find some action verbs or linking verbs.

77 Writer's Companion ■ UNIT 3
Lesson 2 *Staying on Topic*

WRITER'S GRAMMAR: ACTION & LINKING VERBS

Read aloud the Writer's Grammar. Explain that action verbs describe an action (*run, jump,* or *smile*) as in: *Greg kicks the soccer ball.* Then explain that linking verbs connect the subject of the sentence to information about the subject (*is, am,* or *was*). Linking verbs do not describe actions. *Mary's hat is red* is a sentence with a linking verb. Have students suggest other sentences with action verbs and linking verbs; write them on the board.

Evaluate a Speech

OBJECTIVES

- To understand what to look for when evaluating a speech
- To evaluate a speech using a checklist
- To suggest improvements for a speech

Teach/Model Discuss with students why it is important to evaluate writing. Explain that evaluating their work helps to remind writers to see if their writing is the best that it can be and that it met its purpose. Also discuss what to look for in evaluating a speech by using the bulleted items at the top of the page.

Guided Practice Read aloud the directions for Part A. Guide students in writing their responses. Remind them that they will need to look at the Student Model for help. Then explain that using a checklist can help writers evaluate writing by keeping them focused on a few important elements. Point out the checklist in Part B and read aloud the directions for both Parts B and C.

Independent Writing Practice Have students complete Parts B and C independently.

LESSON 2: STAYING ON TOPIC

Name ______________________

Evaluating the Student Model

Evaluate a Speech

When you evaluate a speech, ask yourself these questions:

- Does the writer get the audience's attention? (Look for a strong opinion statement.)
- Does the writer state reasons in a logical order? (Look for an order that makes sense.)
- Does the writer stay on topic? (Look for details that support the opinion.)

A. Reread the Student Model on page 77. Then answer these questions.

1. What is the writer's opinion? Yoga is a great activity that everyone should try.
2. What reasons does the writer give? It helps you relax, get stronger, and stretch.

B. Now evaluate the Student Model. Put a check in the box next to each thing the writer has done well. If you do not think the writer did a good job with something, do not check the box.

- ☐ The writer began with a strong opinion statement.
- ☐ The writer used reasons and details to support the opinion.
- ☐ The writer put reasons and details in a logical order.
- ☐ The detail sentences are on topic.
- ☐ The writer concluded by restating the opinion or calling for action.

C. How do you think the writer could make the speech better? Write your ideas below.

Possible response: The writer should put the reasons and details in a more logical order. She should also conclude by restating the opinion or calling for action.

Writer's Companion ▪ UNIT 3
Lesson 2 *Staying on Topic*

78

See the rubric on page 207 for another way to evaluate the Student Model.

SHARING AND DISCUSSING

Have students share their responses to Parts B and C. Pair students and ask them to tell why they checked a particular item from the checklist. Also have them share their suggestions for improvement from Part C. Then discuss their responses as a class, asking them if they think the writer did a good job overall.

Revise by Rearranging Sentences

OBJECTIVES

- To revise a speech by rearranging its sentences
- To make sure that reasons and details are in a logical order

Teach/Model Tell students that revising writing means making changes to improve it. Explain that there are many different ways to revise writing. In this activity, students will focus on rearranging sentences into the best logical order.

Guided Practice Read aloud the introduction and the example. Explain that reasons and details must follow a logical order, or their speech will not make sense. Tell students to place the most persuasive reason at the end of their speech. Guide students through the example by presenting the original order and asking if the order makes sense. Then read aloud the revision and ask the same question.

Independent Writing Practice After reading aloud both sets of directions, have students complete Parts A and B independently. Invite volunteers to read aloud their revisions.

LESSON 2: STAYING ON TOPIC

Name ______________________

Revising the Student Model

Revise by Rearranging Sentences

When you revise a speech, make sure the reasons and details are in a logical order. Put the most persuasive reason at the end of the speech. Here is an example of how a group of sentences from the Student Model can be rearranged. Think about which order is more effective.

Example Yoga is a great activity that everyone should try. Part of each yoga class is spent just thinking and breathing. It can really relax you.

Revised: Yoga is a great activity that everyone should try. It can really relax you. Part of each yoga class is spent just thinking and breathing.

A. Revise each group of sentences from the Student Model. Change the order. Then put a star next to the order you think works best. Answers may vary.

1. Yoga can help you get stronger. It is a good way to stretch your whole body. It works every muscle you have.

   Possible response: Yoga can help you get stronger. It works every muscle you have. It is a good way to stretch your whole body.

2. It is a good way to stretch your whole body. It works every muscle you have. Even your face and your feet will feel stretched.

   Possible response: It works every muscle you have. It is a good way to stretch your whole body. Even your face and your feet will feel stretched.

B. Revise the speech you wrote on page 76. Rearrange your sentences in a logical order. Put the most persuasive reason at the end. Do your writing on another sheet of paper. Answers will vary. Students should put their reasons on a logical order, with the most persuasive reason at the end.

79 Writer's Companion ■ UNIT 3 Lesson 2 *Staying on Topic*

WRITER'S STRATEGY: REREAD SOFTLY

Tell students that one strategy they can use before presenting their speech is to reread it softly to themselves, listening for any repeated or missing words. They can also check their spelling and punctuation and make sure that they have included all of the important reasons or details. Remind students to take their time as this step will help them improve their speech before presenting it.

Look at Reasons and Examples

OBJECTIVES

- To understand the terms *reason* and *example*
- To identify reasons and examples in literature

Teach/Model After reading aloud the introduction, tell students that when writers give opinions, they must support them with reasons and examples. Explain that a reason tells why something is true, and an example shows that it is true. Read aloud the directions for Part A and the passage from *Centerfield Ballhawk.* As students listen, ask them to focus on how Mr. Mendez supports his opinion.

Guided Practice Read aloud the directions for Part B. For Item 1, ask students what is Mr. Mendez's opinion about José. Does he think José is a good baseball player? Help students find the sentence that states this opinion. Then have volunteers work through Items 2–3, finding a reason and an example that support Mr. Mendez's opinion.

Independent Writing Practice After reading aloud the directions, have students complete Part C independently.

Use a Literature Model

LESSON 3: REASONS AND EXAMPLES

Name ______________________________

Writer's Craft in Literature

Look at Reasons and Examples

When writers want readers to agree with their opinions, they support those opinions with reasons and examples. A **reason** tells why something is true. An **example** shows that it is true.

A. Read the following conversation between José and his father, Mr. Mendez. Notice how José's father supports his opinion.

Literature Model

"And one more thing. Forget about trying to hit like I did, okay? You don't have to. You're a born outfielder, José! You've made catches that I never would have been able to, not in a million years."

José stared at him. "Really? You mean you… don't mind that I can't hit?"

José's father chuckled. "Can't hit'? If you call X belting a grand slam homer not hitting, well, son, we've got to sit down and have a serious talk about the game of baseball! José, you're a born ballhawk, so stop worrying about the hitting and concentrate on your fielding. That's where your team needs you the most."

—from *Centerfield Ballhawk*
by Matt Christopher

B. Find the reasons and examples that support José's father's opinion.

1. Circle the sentence that states José's father's opinion.
2. Underline one reason that supports the opinion.
3. Put an X by one example that supports the opinion.

C. What is Mr. Mendez's opinion of José's playing? Write it in your own words.

Possible response: He thinks José is a great ballplayer.

Writer's Companion ▪ UNIT 3
Lesson 3 *Reasons and Examples* 80

SHARING AND DISCUSSING

Invite students to share their responses to Part C with the class. Have them discuss whether their responses reflect Mr. Mendez's opinion.

Explore Reasons and Examples

OBJECTIVES

- To deepen students' understanding of reasons and examples
- To identify reasons and examples that support an opinion
- To write an example

Teach/Model Review the diagram at the top of the page and discuss the relationship between opinions, reasons, and examples. Make sure students understand that an opinion is what a person thinks about a topic, and that reasons and examples tell or show that an opinion is true.

Guided Practice Read aloud the directions for Part A and the example. Ask students to identify the example. Is it a reason or example? How can they tell? Complete Items 1–2 as a class.

Independent Writing Practice After reading aloud both sets of directions, have students complete Parts B and C independently. Remind students that the introductory diagram can help them as they work.

LESSON 3: REASONS AND EXAMPLES

Name ______________________

A Closer Look at Writer's Craft

Explore Reasons and Examples

When you write an opinion, you can support it with **reasons** and **examples**.

Opinion

Reasons tell why an opinion is true. | **Examples** show that an opinion is true.

A. Read each short passage below. Circle the word that tells if it contains a REASON or an EXAMPLE.

Example Living on a farm is great. You get to take care of all kinds of farm animals.

REASON (circled) EXAMPLE

1. My sister is a wonderful singer. Last year she won a statewide singing contest.

REASON EXAMPLE (circled)

2. Marco is a good friend. I can tell him anything, and he will always listen.

REASON (circled) EXAMPLE

B. Read the passage below. Underline the reason. Circle the example.

My dog is the smartest dog around. He can understand and obey all my commands. When I tell him to count to five, he raises his paw five times.

C. Add another example to the passage above.

Possible response: When I tell him to, he can roll over and play dead.

81

Writer's Companion ■ UNIT 3
Lesson 3 *Reasons and Examples*

Reaching All Learners

BELOW LEVEL

For Parts B and C, pair students with partners or aides and have students verbally describe a new example about "the smartest dog around." Then have partners guide students in writing their ideas into a sentence.

ADVANCED

Encourage students to use adjectives when writing their example for Part C. Remind them to brainstorm a list of appropriate adjectives or to use a dictionary or thesaurus for guidance.

ENGLISH LANGUAGE LEARNERS

Pair less fluent English speakers with more fluent English speakers. Have them discuss possible examples about "the smartest dog around." Then have them look for the correct words using a dictionary or thesaurus.

Use Reasons and Examples

OBJECTIVES

- To prepare to write a persuasive letter
- To write an opinion and a list of supporting reasons and examples

Teach/Model Tell students that before writing a persuasive letter, they should first think about the opinion they want to express. Read aloud the introduction. Then guide students through the example to help them understand the opinion and its supporting reasons and examples.

Guided Practice Read aloud the directions for Part A. Have students write an opinion. Then guide students in filling out their charts of reasons and examples. Also tell students that they may turn back to page 81 to review the meaning of the terms *reason* and *example.*

Independent Writing Practice After reading aloud the directions, have students complete Part B independently. Remind them to use their charts from Part A for assistance.

Name ______________________

Practice with Writer's Craft

Use Reasons and Examples

A persuasive letter states an opinion and supports it with reasons and examples. It tries to persuade readers to do or think something. When you plan a persuasive letter, write your opinion. Then list reasons and examples that support it. Here is how one student planned a persuasive letter.

Example My Opinion: Lin should join the debate team.

Reasons	Examples
We need a good teammate. Lin speaks well in public. Lin is good at doing research.	Our team has lost four of our last five debates. Everyone loved her speech on recycling. She got an A on our last research report.

A. **Think of a friend whom you would like to join a club, team, or activity. Write your opinion. Then list reasons and examples to support your opinion.**

My Opinion: ______________________

Reasons	Examples

B. **Use information from your chart to write a persuasive letter to your friend. Include at least two reasons and one example. Do your writing on another sheet of paper.**
Answers will vary, but should include an opinion, two reasons, and one example.

EXTENDING THE CONCEPT: CLEAR EXAMPLES

Before students complete Part B, have them revise their examples from Part A. Ask: How clear are your examples? Are they easy to imagine? Do vivid pictures quickly come to mind? Guide students in using clear language that readily "paints a picture" of what they want to describe.

The Parts of a Persuasive Letter

OBJECTIVES

- To understand how a persuasive letter is organized
- To analyze a Student Model

Teach/Model Read aloud the introduction and tell students that the Student Model is an example of a persuasive letter written by a third grader. Read aloud and explain the callouts about how to organize and write a persuasive letter. Then read aloud the model. As students listen, ask them to pay attention to how the reasons are organized.

Guided Practice Point out the items following the Student Model. Read aloud Item 1 and guide students in deciding which sentence states Sam's opinion. After answering, complete Item 2 as a class.

Independent Writing Practice Have students do Items 3–4 independently. Then discuss their responses as a class.

Use a Student Model

LESSON 3: REASONS AND EXAMPLES

Name ______________________

Focus on the Writing Form

The Parts of a Persuasive Letter

Below is an example of a **persuasive letter** written by a third grader. As you read, think about how the writer organized the letter. Then answer the questions.

Student Model

DRAFT

97 North Dover Rd.
Afton, MI 49705
October 4, 2006

Dear Lin,

I'm writing to you because I think you would be a great member of our debate team. You speak really well. I saw you give the speech on recycling last month. You were great! X You are also good at telling jokes.

We need you to make our team stronger. Our team has lost four of our last five debates. We need a good speaker to make us a better team.

You are also good at doing research. You got an A on the last research report. You said you liked doing it. Our team really needs you!

Your friend,
Sam

Introduce the topic by writing the heading and the greeting. Write an opening sentence that states your opinion clearly.

State three reasons in a logical order.

Develop your reasons by giving examples.

Conclude by restating your opinion and calling for action.

Finish with a closing and your signature.

1. Which sentence states the writer's opinion? Underline it.
2. What is the strongest reason the writer gives? Put a box around it.
3. Circle one example the writer gives.
4. Which sentence is not supported by reasons and examples? Put an X next to it.

Writer's Grammar
All abbreviations, except for the postal abbreviations of states, need a period at the end. Find a postal abbreviation in the letter.

83 Writer's Companion ■ UNIT 3
Lesson 3 *Reasons and Examples*

WRITER'S GRAMMAR: ABBREVIATIONS

Read aloud the Writer's Grammar. Have students find examples of abbreviations in the Student Model. Are they correctly punctuated? What do they mean? Tell students that postal address abbreviations need to be correct, so that a letter arrives at its correct destination.

Evaluate a Persuasive Letter

OBJECTIVES

- To understand how to evaluate a persuasive letter
- To evaluate a persuasive letter
- To suggest improvements for a persuasive letter

Teach/Model Tell students that when evaluating a persuasive letter, they should look for a strong opinion that is supported with logical reasons and examples. Read the bulleted items at the top of the page to guide students in evaluating the Student Model.

Guided Practice Read aloud the directions for Part A. Have a volunteer read Question 1 aloud. As a class, discuss the ideas and opinions voiced in the letter. Then have a second volunteer read Question 2 aloud. Have students identify the reasons in the letter. Point out the checklist in Part B and explain that it covers the writing skills practiced in this lesson.

Independent Writing Practice After reading aloud both sets of directions, have students complete Parts B and C independently.

Name ______________________________

Evaluating the Student Model

Evaluate a Persuasive Letter

When you evaluate a persuasive letter, ask yourself these questions:

- Does the writer begin with a strong opinion?
  (Look for words that tell what the writer feels or thinks.)
- Does the writer give reasons in a logical order?
  (Look for the strongest reason at the end of the letter.)
- Does the writer support the opinion with examples?
  (Look for words that show that the opinion is true.)

A. Reread the Student Model on page 83. Then answer these questions.

1. What is the writer's opinion?

   The writer thinks that Lin will be a good debater.

2. What reasons does the writer give?

   Lin speaks well, the team needs a good speaker, Lin likes to do research.

B. Now evaluate the Student Model. Put a check in the box next to each thing the writer has done well. If you do not think the writer did a good job with something, do not check the box.

- ☐ The writer began with a strong opinion.
- ☐ The writer supported the opinion with reasons that tell why the opinion is true.
- ☐ The writer used examples that show why the opinion is true.
- ☐ The writer put reasons from least important to most important.
- ☐ The writer ended by restating the opinion and calling for action.

C. How do you think the writer could make the end of the letter better? Write your ideas below.

Possible response: The writer could have asked Lin to join the club, calling her to action.

See the rubric on page 207 for another way to evaluate the Student Model.

SHARING AND DISCUSSING

Pair students and have them share their responses from Part C. Also ask them to give one new reason and example to better convince Lin to join the debate team. Ask: If you were Lin, would these new reasons easily convince you to join? Discuss student responses as a class.

Revise by Rearranging Reasons

OBJECTIVES

- To revise a persuasive letter by rearranging reasons

Teach/Model Explain that revising allows writers the chance to improve their writing. Tell students that this activity will focus on rearranging reasons.

Guided Practice Read aloud the introduction. Explain that when revising a persuasive letter, writers should make sure that their reasons are as strong as they can be. Tell students that, in this activity, they will be revising the Student Model and their writing by rearranging reasons from the weakest to the strongest. Then read aloud the directions for Part A and the Word Bank before guiding students in rearranging reasons.

Independent Writing Practice After reading aloud the directions, have students do Part B independently. Invite volunteers to share their revisions with the class.

LESSON 3: REASONS AND EXAMPLES

Name ______________________

Revising the Student Model

Revise by Rearranging Reasons

When you revise a persuasive letter, you should start with the weakest reason and move to the strongest reason. One thing the writer could have done better is rearrange his reasons from weakest to strongest.

A. Rearrange the body of the Student Model. Change the order of the reasons from weakest to strongest. Use the Word Bank to help you.

I'm writing to you because I think you would be a great member of our debate team. You speak really well. I saw you give the speech on recycling last month. You were great! You are also good at telling jokes.

We need you to make our team stronger. Our team has lost four of our last five debates. We need a good speaker to make us a better team.

You are also good at doing research. You got an A on the last research report. You said you liked doing it.

Word Bank
first
next
most important

Order may vary.

I'm writing to you because I think you would be a great member of our debate team. First, you are good at doing research. You got an A on the last research report. You said you liked doing it.

Next, you speak really well. I saw you give the speech on recycling last month. You were great! You are also good at telling jokes.

Most important, we need you to make our team stronger. Our team has lost four of our last five debates. We need a good speaker to make us a better team.

B. Revise the letter you wrote on page 82. Put the reasons in order from weakest to strongest. Write your letter on a separate sheet of paper. Answers will vary. Students should put reasons in order from weakest to strongest.

85 Writer's Companion ■ UNIT 3
Lesson 3 *Reasons and Examples*

WRITER'S STRATEGY: TALKING OUT IDEAS

Pair students. Tell them that a helpful way to revise is to talk out your ideas. Before completing Part B, have students talk out with their partners how they might rearrange the reasons in their letter from page 82. Then discuss as a class whether this activity was helpful.

Review Ideas

OBJECTIVES

- To review the terms *opinion, stay on topic,* and *reasons and examples*
- To identify opinion and reasons and examples in literature

Teach/Model After reading aloud the introduction, tell students that when writers give opinions, they need to provide supporting reasons and examples. Remind students that to stay on topic means to include only material that relates to the chosen topic. Point out that it is also important for the detail sentences to stay on topic. Read aloud the directions for Part A and the passage from *Ramona Forever*. As students listen, ask them to think about Ramona's opinion.

Guided Practice Have a volunteer read aloud the directions for Part B. Then have students read the passage silently to themselves. For Item 1, ask, what is Ramona's opinion about Roberta? Then have students complete Item 2 by identifying reasons from the passage that supports this opinion.

Independent Writing Practice After reading aloud the directions, have students complete Part C independently.

Use a Literature Model

LESSON 4: REVIEW IDEAS

Name ______________________________

Writer's Craft in Literature

Review Ideas

When you write an opinion, you should support it with reasons and examples. Your details should all stay on topic.

A. Below is a passage from *Ramona Forever*. As you read it, think about Ramona's opinion.

Literature Model

Ramona sat back and buckled her seat belt. She had once looked like Roberta. Amazing! She had once been that tiny, but she had grown, her hair had calmed down when she remembered to comb it, and she had learned to use her eyes and hands. "You know what I think?" she asked, and did not wait for an answer. "I think it is hard work to be a baby."

—from *Ramona Forever*
by Beverly Cleary

B. Reread the passage to find the opinion and reasons that support it.

1. Underline Ramona's opinion.
2. Circle the reasons Ramona gives.

C. Write a new example that supports Ramona's opinion.

Possible response: "My baby sister took a long time to learn how to talk well."

____

Writer's Companion ■ UNIT 3
Lesson 4 *Review Ideas*

86

SHARING AND DISCUSSING

Place students in small groups and have them discuss their responses from Part C. What new examples did they come up with about being a baby? Then ask volunteers to share their examples with the class.

Review Ideas

OBJECTIVES

- To review the elements of an *opinion*
- To practice writing an opinion and reasons and examples that stay on topic

Teach/Model Remind students that writers express opinions with supporting reasons and examples. Review the diagram at the top of the page and discuss how writers use reasons and examples while staying on topic.

Guided Practice Read aloud both the directions for Part A and the passage. Ask students why is it "important to be prepared for a fire at home?" What reasons and examples in the passage support this opinion? Then read aloud the directions for Part B and review the example as a class.

Independent Writing Practice Have students complete Parts B and C independently.

LESSON 4: REVIEW IDEAS

Name ______________________

A Closer Look at Writer's Craft

Review Ideas

Writers often express opinions. Then they give reasons and examples to support these opinions.

Opinion

Reasons tell why the opinion is true. | Reasons and examples should **stay on topic**. | **Examples** show that the opinion is true.

A. **Read the beginning of this article. Look for reasons and examples that support the opinion.**

It is important to be prepared for a fire at home. Floods can also be dangerous. If you and your family have a fire safety plan, you will be safer in a fire. The Pérez family's house burned down last month. They followed their fire safety plan. No one was hurt. A fire safety plan can also help firefighters.

B. **Write the information.**

Example **Write a reason:** If you and your family have a fire safety plan, you will be safer in a fire.

1. Write an opinion: "It is important to be prepared for a fire at home."

2. Write a reason: "A fire safety plan can also help firefighters."

3. Write an example: "The Pérez family's house burned down last month. They followed their fire safety plan. No one was hurt."

C. **Write the sentence that is not on topic.**

"Floods can also be dangerous."

87 Writer's Companion ▪ UNIT 3 Lesson 4 *Review Ideas*

EXTENDING THE CONCEPT: OFF-TOPIC SENTENCE

Convert the off-topic sentence from Part C into an opinion—*It is important to be prepared for a flood.* Have students write two reasons and examples supporting this opinion.

Review Ideas

OBJECTIVES

- To form an opinion about a book
- To write reasons and examples supporting an opinion for a book review

Teach/Model Review how to plan for a book review. Tell students that the first step in writing a book review is forming an opinion about the book. Read aloud the introduction.

Guided Practice Review the example as a class. Have several volunteers read aloud the given opinion and reasons. Ask students to think about the reasons. Do they support the opinion? Then complete and review Items 1–2 as a class.

Independent Writing Practice After reading aloud the directions, have students complete Part B independently. Invite volunteers to share their examples aloud.

LESSON 4: REVIEW IDEAS

Name ____________________

Practice with Writer's Craft

Review Ideas

When you plan a review of a book, you decide if you like the book or not. Then you give reasons and examples for your opinion. Read the example in the box below. It shows how a third grader began to think about a book review.

My Opinion of *Ramona Forever*:	I loved the book.
Reason 1:	Ramona is a funny character.
Reason 2:	Ramona feels the way I do about many things.
Reason 3:	Ramona learns from her mistakes.

A. **1. How does the writer feel about *Ramona Forever*?**

____

____

2. Which of the reasons would be most important to you? Explain why.

____

____

B. Write a new example to support one of the reasons from the chart. Tell which reason you are supporting.

____

____

____

____

____

____

Writer's Companion ▪ UNIT 3
Lesson 4 *Review Ideas* 88

Reaching All Learners

BELOW LEVEL

Before completing Part B, have students work in small groups and talk about possible new examples to add to the chart. After talking out their ideas, have students write a couple of new examples. Encourage students to review their examples to make sure that they support the selected reason.

ADVANCED

Encourage students to write a new opinion about *Ramona Forever*. Have students brainstorm ideas for a new opinion with a partner. After forming their opinion, have partners write a couple of reasons and examples to support the new opinion.

ENGLISH LANGUAGE LEARNERS

Pair less fluent English speakers and more fluent English speakers. Have students brainstorm possible new examples for Part B and then have them write them down. Encourage students to review their examples to make sure that they support the selected reason.

The Parts of a Review

OBJECTIVES

- To understand how a review is organized
- To analyze a Student Model

Teach/Model Read aloud the introduction. Then review the callouts about how to organize and write a book review. Read aloud the Student Model. As students listen, ask them to think about how the opinion is expressed. Also have students think about how the writer used reasons and examples to support her opinion.

Guided Practice Point out the items following the Student Model. Have a volunteer read aloud Item 1. Complete it as a class. Remind students that they need to refer to the Student Model to find their answers.

Independent Writing Practice Have students complete Items 2–3 independently. Then discuss their responses as a class.

Use a Student Model

LESSON 4: REVIEW IDEAS

Name ____________________

Focus on the Writing Form

The Parts of a Review

In this unit, you have learned how writers develop their work when they write an opinion. Below is a first draft of a review written by a third grader. As you read, think about how the student organized the review. Then answer the questions.

Student Model

DRAFT

Ramona Forever
by Gina

I loved the book *Ramona Forever* by Beverly Cleary. It is the story of Ramona Quimby and her family. She has a sister, Beezus. Her mother is about to have a baby. Ramona isn't too happy about the baby. She has always been the baby of the family. She is afraid the baby will take her place. Babies are often troublesome.

Ramona feels the way I do about many things. She has mixed-up feelings about her family and friends. She has a terrible argument with her sister. Just like a real person.

Ramona is a funny character. She is always making mistakes and getting in trouble. Still, she learns from her mistakes.

I thought *Ramona Forever* was a wonderful book about a real-life girl.

Introduce the review by stating your opinion.

Support your opinion with reasons and examples.

Provide details that are on topic.

End by restating your opinion and calling readers to action.

1. Underline the sentence that tells an opinion.
2. Underline a reason that supports the opinion. Possible response given.
3. Draw a box around an example that tells about the reason. Possible response given.

Writer's Grammar
Remind students that a pronoun takes the place of a noun. Explain that a subject pronoun is one that can be used as the subject of a sentence: *I, you, he, she, it, we,* and *they* are subject pronouns. Find a subject pronoun in the Student Model.

89

Writer's Companion ■ UNIT 3
Lesson 4 *Review Ideas*

WRITER'S GRAMMAR: SUBJECT PRONOUNS

Have several volunteers read aloud the Writer's Grammar. Help students identify subject pronouns in the model. Then have them suggest their own sentences containing subject pronouns. Write and review them on the board.

Evaluate a Review

OBJECTIVES

- To evaluate a review that received a score of 4 off of a 4-point rubric (*excellent*)
- To see how a 4-point rubric is used to evaluate a review

Teach/Model Explain that a *rubric* is a scoring chart often used to evaluate writing. Point out the rubric on the following page. Tell students that they will learn how writing is scored using a rubric on the following two pages.

Guided Practice Read aloud the directions for Part A and the Student Model, or have students take turns reading aloud. As students listen, ask them why do they think *The Saturdays* was scored as an "excellent" book review. Then read aloud and review the teacher's comments. Ask students to identify the opinion and a supporting reason and example within the model.

Use a Student Model

LESSON 4: REVIEW IDEAS

Name ______________________

Evaluating the Student Model

Evaluate a Review

A. Two students were asked to write a review. The review below got a score of 4. When using a 4-point rubric, a score of 4 means "excellent." Read the review and the teacher comments that go with it. Find out why this is a success.

Student Model

DRAFT

The Saturdays
by Veronica

The Saturdays by Elizabeth Enright is an amazing book that is really fun to read. It is the story of the four Melendy children, Mona, Rush, Randy, and Oliver. One rainy Saturday, they are very bored. Randy has a great idea. They will form a club, the Saturdays. They will put all their allowance money together. Each Saturday, one of them will take all the money and have an adventure.

Randy goes first because the club was her idea. She goes to the ballet. There she meets Mrs. Oliphant. She is an old friend of the family. Mrs. Oliphant takes Randy to tea and tells her thrilling stories of her own childhood.

Mona goes next. She gets her long, long hair chopped off. Then Rush goes to the opera. He loves music, so the opera is very special to him. Last, it is Oliver's turn. He is only six, so he is not allowed to go out by himself. He sneaks out anyway and goes to the circus. Then he gets lost. A policeman brings him home on a horse.

The author makes these characters come alive. I felt like I knew them all. Their adventures are exciting to read about. I wished I could go along. You'll find that reading about the Melendys' adventures is almost as much fun as having them yourself!

Good – you state your opinion strongly here.

These examples are very nice!

Good word choice.

You use different sentence structures here to keep the review interesting.

You give some excellent reasons to read the book.

Nice – you restate your opinion and call your readers to action!

WRITER'S STRATEGY: PROOFREADERS

Tell students that when they are revising their writing, they should have a proofreader. They can ask a classmate, a family member, or friend to read their writing. Explain that they should ask their proofreader to check for logic, misspellings, and grammatical errors. Tell students that having a proofreader will help improve their writing.

Evaluate a Review

OBJECTIVES

- To review how to evaluate a review that received a score of 2 off of a 4-point rubric
- To see how a 4-point rubric is used to evaluate a review

Teach/Model Read aloud the directions for Part B and tell students that the following Student Model received a 2-point score on a 4-point rubric. This means that this review "needs improvement." Read aloud the Student Model. As students listen, ask them to think about why this review needs improvement.

Guided Practice Before reading aloud the teacher's comments, ask students to share their thoughts about the strengths and weaknesses of this review. Then have volunteers read aloud the teacher's comments.

Independent Writing Practice After reading aloud the directions, explain how to use a 4-point rubric. Then have students complete Part C independently.

Use a Student Model

LESSON 4: REVIEW IDEAS

Name ____________________

Evaluating the Student Model

B. This paragraph got a score of 2. Why did it get a low score?

Student Model

DRAFT

Miss Rumphius
by Marla

I liked the book *Miss Rumphius* by Barbara Cooney. Miss Rumphius wants to see the world. She travels all over and sees beautiful things. Then she falls off a camel. She hurts her back. She goes home to her little house. She wants to make her yard beautiful. She plants some lupine flowers. They grow and bloom.

Miss Rumphius plants lupines farther and farther away. She scatters seeds all around. She is called the Lupine Lady. The flowers bloom everywhere. Miss Rumphius has helped to make the world beautiful.

Good – you state your opinion here.

These examples tell about the book.

You need to include some reasons to support your opinion.

You should vary your sentence structures to make your writing more interesting.

Here, you need to restate your opinion and call your readers to action.

C. What score would you give the student's story? Put a number on each line.

	4	3	2	1
Ideas ______	☐ Writer gives a strong opinion with reasons and examples.	☐ Writer gives an opinion with some reasons or examples.	☐ Writer only gives an opinion.	☐ Writer does not give an opinion, reasons, or examples.
Words and Sentences ______	☐ Writer uses vivid words and details. Sentences vary.	☐ Writer uses some vivid details and varied sentences.	☐ Writer uses few vivid details. Sentences do not vary.	☐ Writer uses unclear words. Sentences do not vary.
Organization ______	☐ Student presents information in logical, interesting sequence, which audience can follow.	☐ Student presents information in logical, sequence, which audience can follow.	☐ Audience has difficulty following presentation because student jumps around.	☐ Audience cannot understand presentation because there is no sequence of information.

91

Writer's Companion ▪ UNIT 3
Lesson 4 *Review Ideas*

EXTENDING THE CONCEPT: REVIEWING OPINION

Review with students that an opinion tells how you feel or what you think about something. Have them write an opinion for the Student Model entitled *Miss Rumphius*. Explain that they should state what they think of the book and give reasons and examples to support that opinion.

Extended Writing/Test Prep

OBJECTIVES

- To become familiar with test assessment writing prompts
- To select a writing prompt and begin the prewriting process for a longer written work

Teach/Model Tell students that the following two pages will give them the opportunity to practice writing. Explain that they will choose from three possible writing prompts. Read the directions for Part A and each writing prompt aloud. Guide students in choosing a writing activity.

Guided Practice Review the writing forms in Item 2. Point out that if they choose Item 2 students might want to review the lesson material about that particular type of writing. Read aloud the Student Model callouts per each writing form.

- A Persuasive Paragraph: Lesson 1, Page 71
- A Speech: Lesson 2, Page 77
- A Persuasive Letter: Lesson 3, Page 83
- A Review: Lesson 4, Page 89

Then have students begin the prewriting process by completing Part B independently.

LESSON 4: REVIEW IDEAS

Name ______________________

Extended Writing/Test Prep

Extended Writing/Test Prep

On the last two pages of this lesson, you will use what you have learned to write a longer written work.

A. Read the three choices below. Put a star by the writing activity you would like to do.

1. Respond to a Writing Prompt

   Writing Situation: Think about a story or book you have read. Maybe you really liked it. Maybe there were some things about it you did not like.

   Directions for Writing: Now write a review of the story or book. Tell your opinion. Include reasons and examples.

2. Choose one of the pieces of writing you started in this unit:

   - a persuasive paragraph (page 70)
   - a speech (page 76)
   - a persuasive letter (page 82)

   Expand your beginning into a complete piece of writing. Use what you have learned about writing opinions.

3. Choose a topic you would like to write about. You may write a persuasive paragraph, a speech, or a persuasive letter. Use reasons and examples to support your opinion.

B. Use the space below and on the next page to plan your writing.

TOPIC: ______________________

WRITING FORM: ______________________

HOW WILL I ORGANIZE MY WRITING: ______________________

____

____

Writer's Companion ▪ UNIT 3
Lesson 4 *Review Ideas* 92

SHARING AND DISCUSSING

Place students in small groups to discuss their writing activity. Have each student explain what they will write about, which writing form they will use, and how they will organize their writing. Encourage partners to give suggestions that may help in the prewriting process. Then discuss as a class.

Extended Writing/Test Prep

OBJECTIVES

- To use a graphic organizer to plan a piece of writing
- To use the steps in the writing process to complete a longer piece of writing independently

LESSON 4: REVIEW IDEAS

Name ______________________________

Extended Writing/Test Prep

C. In the space below, draw a graphic organizer that will help you plan your writing. Fill in the graphic organizer. Write additional notes on the lines below.

Notes

____

____

____

____

____

____

____

____

____

D. Write your paragraph, speech, or letter on another sheet of paper.

93 Writer's Companion ■ UNIT 3 Lesson 4 *Review Ideas*

Teach/Model Read aloud the directions for Part C. Then explain to students that they will select a graphic organizer to plan their writing. Help each student select an appropriate graphic organizer from the back pages of this book. On the board, model how to complete a graphic organizer before students complete their own.

Guided Practice Help students complete their graphic organizers. Remind them that a graphic organizer is a general guide to use as they write. Explain that they do not need to use everything that they include, and they can add other details later on in their writing.

Independent Writing Practice: Draft, Revise, Publish Have students do their writing on another sheet of paper. Remind students that, after their first drafts, they will need to revise their writing. Then ask students to think about how they want to publish their work. Tell students that they may choose to create illustrated front and back covers or present their writing in an oral presentation.

WRITER'S STRATEGY: PUBLISHING OPTIONS

Pair students. Have students talk about how they can publish their writing. Do they want to create a book, class anthology, a billboard, or an oral presentation? Have them brainstorm ideas to help them decide how to publish their writing. Then ask volunteers to share their ideas with the class.

Answering Multiple-Choice Questions

OBJECTIVES

- To practice answering multiple-choice questions using an Answer Sheet
- To make a plural noun possessive by adding an apostrophe (') after the *s*

Teach/Model Explain that some writing tests include multiple-choice questions. Tell students that this lesson will let them practice answering multiple-choice questions using an Answer Sheet. Read aloud the introduction, test tip, and directions. Complete Question 1 as a class. Make sure to read aloud the question and the three possible answers before selecting a response. Tell students that they should always read all possible answers before selecting their final response.

Guided Practice Have a volunteer read aloud Question 2 and then have students read the possible choices silently to themselves. Ask students which sentence has correct capitalization. Have students mark the correct answer on their Answer Sheets.

Independent Writing Practice Review the test tip and tell students that it will help them complete Question 3. Have students complete the last two items independently. Remind them to check if they filled in their Answer Sheets correctly.

Name ______________________________

Writing Test Practice

Answering Multiple-Choice Questions

This lesson will help you answer different kinds of multiple-choice questions often found in writing tests. Read each question. Mark your answer on the Answer Sheet.

A. For some multiple-choice questions, you will have to read three or four sentences and decide which sentence is written correctly. Read the test tip. Then practice answering this type of test question.

1. Which sentence below is correct?
   - A The girls' bikes were piled in a heap.
   - B The girls bikes were piled in a heap.
   - C The girls bikes' were piled in a heap.
2. In which sentence below is all **capitalization** correct?
   - F I opened my mouth so Dr. parsons could look inside.
   - G I opened my mouth so Dr. Parsons could look inside.
   - H I opened my mouth so dr. Parsons could look inside.
3. Which sentence below is correct?
   - A The cars horns all honked at once.
   - B The cars horns' all honked at once.
   - C The cars' horns all honked at once.
4. Which sentence below is correct?
   - F Mr. and Mrs. Sami live next door.
   - G Mr. and mrs. sami live next door.
   - H Mr and Mrs Sami live next door.

Test Tip: To make a plural noun possessive, add an apostrophe (') after the *s*. (Example: my friends' jackets.)

Answer all test questions on this Answer Sheet.

1.	**(A)**	(B)	(C)	3.	(A)	(B)	**(C)**
2.	(F)	**(G)**	(H)	4.	**(F)**	(G)	(H)

SHARING AND DISCUSSING

Pair students and have them compare and discuss their responses for the last two items. Ask them to share their reasons for choosing their answers for each item.

Answering Multiple-Choice Questions (cont.)

OBJECTIVES

- To practice reading a passage before answering questions
- To become familiar with reading a passage in which sentences are numbered

Teach/Model Explain to students that, in some multiple-choice tests, they will have to read a passage and then choose the best way to correct a sentence. Read aloud the introduction, directions, and the passage. Then complete Question 1 as a class and read aloud the test tip about a personal narrative.

Guided Practice Guide students in completing Question 2. Tell them to follow the same steps they did in answering Question 1. Remind students to read the question and all possible answers before selecting their final response.

Independent Writing Practice Have students complete Question 3 independently.

LESSON 5: WRITING TEST PRACTICE

Name ______________________________

Writing Test Practice

B. For some multiple-choice questions, you will have to read a passage and then decide the best way to revise or correct a sentence. Read the test tip. Then practice answering this kind of question.

DIRECTIONS

Read the introduction and the passage that follows. Then read each question and fill in the circle next to the correct answer.

Cilla is in the third grade. She wrote this personal narrative. She wants you to help her revise and edit the narrative. Read Cilla's narrative and think about the changes she should make. Then answer the questions that follow.

The Flood

(1) It was the fourth rainy day in a row. (2) The river behind the house was too full. (3) It just kept raining. (4) I and my brothers were worried. (5) Mom and Dad rushed we into the car. (6) We came back three days later. (7) My brothers shouts told us what happened. (8) The river had come into our house. (9) There was mud in every room!

1. What is the **BEST** way to revise sentence 4?
   - (A) I and my brothers was worried.
   - **(B)** My brothers and I were worried.
   - (C) I and my brothers were worry.
   - (D) No revision is needed.

2. What change, if any, should be made in sentence 5?
   - **(F)** Change *we* to **us**
   - (G) Change *Dad* to **dad**
   - (H) Change *rushed* to **rushing**
   - (J) Make no change

3. What change, if any, should be made in sentence 7?
   - (A) Change *shouts* to **shouted**
   - (B) Change *my* to **me**
   - **(C)** Change *brothers* to **brothers'**
   - (D) Make no change

Test Tip:
When listing people in a personal narrative, list yourself last.

INCORRECT:
I and he went outside.

CORRECT:
He and I went outside.

95 Writer's Companion ■ UNIT 3
Lesson 5 *Writing Test Practice*

ASSESSING STUDENT RESPONSES

If a student answers an item incorrectly, try to find out why. Be aware that an incorrect response may be the result of misunderstanding the directions or not reading all of the choices before selecting one. When trying to decide why a student answered incorrectly, ask him or her to show you their process in choosing the answer for a particular item. Then provide additional instruction and practice as needed.

Answering Multiple-Choice Questions (cont.)

OBJECTIVES

- To practice answering multiple-choice questions using a graphic organizer
- To decide if a statement is on or off topic

Teach/Model Explain that some multiple-choice tests are based on a graphic organizer. Tell students to refer to the graphic organizer before answering any questions. Read aloud the directions and the test tip before reviewing the graphic organizer as a class.

Guided Practice Complete Question 1 as a class. Ask, what kind of paper is Samara planning to write? Then review all possible answers before having the class select the correct answer.

Independent Writing Practice Have students complete Question 2 independently.

Name ____________________

Writing Test Practice

C. **For some writing tests, you will have to answer questions based on a graphic organizer. Samara made the plan to organize ideas for a paper. Use her plan to answer questions 1 and 2 on the Answer Sheet.**

Story/Movie/Show: Mr. Popper's Penguins	**Audience:** My classmates
Opinion: Mr. Popper's Penguins is a great book.	

Reason 1: I love to read. **Details:**
Reason 2: The action in the story is very exciting. **Details:** Mr. Popper is in a car chase to get ice for his penguins.
Reason 3: **Details:** Waddling penguins are so cute.
Opinion Restated/Action Requested: You should read this book!

1. Based on the information in Samara's Writing Plan, what kind of paper is she planning to write?
   - A a story
   - B a review
   - C a how-to essay
   - D a journal/diary entry
2. Which statement below is off topic and should be taken out of Samara's Writing Plan?
   - F *Mr. Popper's Penguins* is a great book.
   - G The action in the story is very exciting.
   - H I love to read.
   - I You should read this book!

Test Tip: To decide if a statement is off topic, ask yourself what it tells you about the topic of a particular piece of writing. If it does not add information, it is off topic.

Answer all test questions on this Answer Sheet.

1. Ⓐ ● Ⓒ Ⓓ   2. Ⓕ Ⓖ ● Ⓘ

Reaching All Learners

BELOW LEVEL

Pair students with partners or aides and have them restate in their own words what each question asks them to do. Then ask students to verbally state the steps they need to take when answering each question successfully.

ADVANCED

Encourage students to review the test tip on their own and to use this information to verify their answers. Have them compare their responses with those of a partner.

ENGLISH LANGUAGE LEARNERS

Have students go through the graphic organizer, step by step, making sure they understand the topic, opinion, and details. Remind students that they need to use the graphic organizer to answer all questions.

Answering Multiple-Choice Questions (cont.)

OBJECTIVES

- To practice answering multiple-choice questions
- To select the sentence that is written correctly

Teach/Model Read aloud the directions for Part D and the test tip. Then have several volunteers read aloud Question 1 and its possible answers. Complete Question 1 as a class.

Guided Practice Continue guiding students as they complete Question 2 by following the same process used in answering Question 1.

Independent Writing Practice Have students complete the last item independently. Remind students that the test tip may provide helpful information.

LESSON 5: WRITING TEST PRACTICE

Name ____________________

Writing Test Practice

D. **For some multiple-choice questions, you will choose which sentence or group of words is written correctly. Read the test tip. Read each question and mark your answer on the Answer Sheet.**

1. Which sentence is written correctly?
   - A The girls was so mad that they could hardly talk.
   - B The girl was so mad that she could hardly talk.
   - C The girl was so mad that her could hardly talk.
   - D The girl was so mad that them could hardly talk.
2. Which sentence is written correctly?
   - A The singers' voices were clear and high.
   - B The singers voices' were clear and high.
   - C The singer's voice's were clear and high.
   - D The singer's voices were clear and high.
3. Which address is written correctly?
   - A 2041 South Millburn ave.
Carmel, CA. 90041
   - B 2041 South Millburn Ave
Carmel, ca 90041
   - C 2041 South Millburn Ave.
Carmel, CA 90041
   - D 2041 South Millburn ave
Carmel, ca 90041

Test Tip: All postal abbreviations are written in full capital letters.

Answer all test questions on this Answer Sheet.

1. Ⓐ **Ⓑ** Ⓒ Ⓓ
2. **Ⓐ** Ⓑ Ⓒ Ⓓ
3. Ⓐ Ⓑ **Ⓒ** Ⓓ

97

Writer's Companion ▪ UNIT 3
Lesson 5 *Writing Test Practice*

USING ACADEMIC LANGUAGE

Most tests contain academic language. Academic language includes direction words such as *identify* or *analyze.* Before asking students to complete any test item, discuss the meaning of these terms: *graphic organizer, story, audience, opinion,* and *reason.*

Look at Combining Sentences

OBJECTIVES

- To understand the term *combining sentences*
- To identify combined sentences in literature

Teach/Model Using the introduction, tell students that short, choppy sentences may be uninteresting to read. Explain that by combining short sentences, writing will be more interesting and easier to understand. Tell students that writers use conjunctions, such as *and* or *but*, to combine sentences.

Read aloud the directions for Part A and the passage from *Papa Tells Chita a Story*. As students listen, ask them to notice how the writer uses conjunctions to combine sentences.

Guided Practice Ask students to find a long sentence that might have been combined. Then guide students in completing Part B.

Independent Writing Practice After a volunteer reads aloud the directions, have students complete Part C independently.

Use a Literature Model

Name ______________________________

Look at Combining Sentences

Short, choppy sentences often make writing uninteresting to read. By **combining** subjects, predicates, and whole sentences, your writing will become more interesting and easier to understand. Use the **conjunctions** *and* or *but* when you combine parts of sentences or whole sentences.

A. Read the following model. Notice how the writer uses conjunctions to combine sentences.

Literature Model

"You're right, Chita! The snake was brown and round and as long as this living room. It raised up its head, squinted its beady eyes, and squinched toward us. But Majestic reared his front legs and zigzagged through the field, while I waved my sword and shouted strong words. We confused that snake, and it snake-snaked away in another direction."

—from *Papa Tells Chita a Story* by Elizabeth Fitzgerald Howard

B. Identify the combined sentences.

1. Underline sentences that might have been combined.
2. Circle one or more conjunctions in each sentence you underlined.

C. Choose one of the sentences you underlined. On the lines below, write two short sentences that might have been combined.

Possible response: The snake was brown and round. The snake was as long as this living room.

SHARING AND DISCUSSING

Have students share their Part C responses with the class. Discuss how short sentences change the rhythm, or flow, of the writing. Ask students which sentences they prefer—short or long sentences—and why. Remind students that the goal is clear and interesting writing.

Explore Combining Sentences

OBJECTIVES

- To deepen students' understanding of how to combine short sentences by using conjunctions
- To combine short sentences by using *and* or *but*

Teach/Model Use the diagram at the top of the page to help students recognize how conjunctions *(and* or *but)* are used in combining short sentences. Explain that *and* combines similarities, while *but* combines differences.

Guided Practice Read the directions for Part A aloud. Write the two sentences from Item 1 on the board. Ask students if Sonia and Robert share something in common. Since the answer is *yes*, ask which conjunction is needed to combine these sentences. After identifying *and*, write the newly combined sentence on the board.

Independent Writing Practice Have students complete the rest of Part A and all of Part B independently. Before beginning Part B, read aloud its directions and see Extending the Concept.

LESSON 1: COMBINING SENTENCES

Name ______________________

A Closer Look at Writer's Craft

Explore Combining Sentences

Combine short sentences with *and* or *but* to make longer, smoother sentences.

Short Sentences
- Suzi likes tacos. Bret likes tacos.
- Suzi likes pizza. Bret likes subs.
- Suzi eats pizza. Suzi drinks milk.

→ *and* or *but* →

Combined Sentences
- Suzi *and* Bret like tacos.
- Suzi likes pizza, *but* Bret likes subs.
- Suzi eats pizza *and* drinks milk.

A. Use a conjunction to combine each pair of sentences. Write the combined sentences on the lines.

1. Sonia has skates. Robert has skates.
   Sonia and Robert have skates.
2. Lucas saw smoke. Lucas called for help.
   Lucas saw smoke and called for help.
3. Roni wants to go to the movies. Roni must do her homework.
   Roni wants to go to the movies, but she must do her homework.

B. Read this combined sentence from *Papa Tells Chita a Story.* Write three smaller sentences that might have been combined to make the longer sentence.

But now Papa is resting and reading and waiting for Chita.

1. But now Papa is resting.
2. Papa is reading.
3. Papa is waiting for Chita.

99

Writer's Companion ▪ UNIT 4
Lesson 1 *Combining Sentences*

EXTENDING THE CONCEPT: COMBINING SENTENCES

Before students complete Part B, remind them that when they combine sentences they add a conjunction. However, when they change long sentences into short sentences, students will have to delete the conjunction or conjunctions. After they are finished, invite students to read aloud their new sentences. Compare the new sentences with the original sentence.

Use Combining Sentences

OBJECTIVES

- To prepare to write by choosing a topic and writing similar details
- To practice comparing and contrasting details by using words such as *both, alike,* and *similar*

Teach/Model Tell students that they will be writing a paragraph that compares two things. Ask students to think about two people they know and to keep these people in mind while reading the example. Then read aloud the introduction and the example.

Guided Practice Read the directions for Part A aloud. Have students write the names of two people they know and then complete the chart. Guide students to include details about each person. Ask questions such as, *What do they like to do? Do they have a pet?*, and so on.

Independent Writing Practice Have students complete Part B independently.

LESSON 1: COMBINING SENTENCES

Name ______________________

Practice with Writer's Craft

Use Combining Sentences

Before you compare any two things or people, think about how they are alike. Combining your sentences can help you compare. Words like *both, alike,* and *similar,* as well as conjunctions, can help you combine subjects, predicates, and sentences. Here is how one student started to think about comparing two of her friends.

Things about Janey	Things about Jessie	Things about Both
Janey is smart.	Jessie is smart.	Janey and Jessie are both smart.
Janey wears jeans and T-shirts.	Jessie wears jeans and T-shirts.	They dress alike in jeans and T-shirts.
Janey plays piano.	Jessie plays flute.	Janey and Jessie are musicians.
Janey loves to read.	Jessie writes stories and poems.	Janey and Jessie like language arts.

A. Think about two people you know. Write their names on the lines. Then complete the chart.

Things about ________	Things about ________	Things about Both

B. Use information from your chart to write a short paragraph comparing two people. Use another sheet of paper.

Writer's Companion ▪ UNIT 4
Lesson 1 *Combining Sentences* 100

Reaching All Learners

BELOW LEVEL

Have students divide a sheet of paper into two columns and label each column with a person's name. Have students draw a picture of each person. Guide them to think of words and phrases that describe each person. Then fill in the chart with these words and phrases.

ADVANCED

Encourage students writing above grade level to use a thesaurus to select synonyms to compare each person. Remind them to look for specific and interesting words.

ENGLISH LANGUAGE LEARNERS

Pair less fluent English speakers with more fluent English speakers. Have students discuss things they know about the people they chose. Then have them help each other identify appropriate English words and phrases to include in the chart.

The Parts of a Paragraph that Compares

OBJECTIVES

- To understand how a paragraph that compares is organized
- To analyze a Student Model

Teach/Model Tell students that the model on this page is a first draft of a third grader's paragraph that compares. Read the introduction and the Student Model aloud. As students listen, ask them to pay attention to the details that compare Janey and Jessie.

Guided Practice Point out the items below the Student Model. Read the first item aloud. Guide students in identifying and underlining the topic sentence. Explain that while a topic sentence is most often found at the beginning of a paragraph, it can also be found in the middle, or at the end of a paragraph.

Independent Writing Practice Have students complete Items 2–4 independently. Invite volunteers to share their responses with the class.

Use a Student Model

LESSON 1: COMBINING SENTENCES

Name ____________________

Focus on the Writing Form

The Parts of a Paragraph that Compares

A **paragraph that compares** begins with a topic sentence. The other sentences give examples and details that support the topic. Below is a draft by a third grader. As you read, look for examples and details that compare the two people.

Student Model

DRAFT

My Two Best Friends
by Samantha

My two best friends Janey and Jessie are a lot alike. People get them mixed up. First of all, they look alike. Janey and Jessie have straight dark hair. Janey and Jessie have bangs. They also dress alike. Janey and Jessie wear the same kind of jeans. They wear they same kind of sneakers. They are both smart. Also, they have similar interests. Janey likes to read, but Jessie writes stories and poems. They both like music. Janey plays piano. Jessie plays flute. Also, Janey's name starts with a J. Jessie's name starts with a J. They are my best friends in the whole entire world.

Begin with a topic sentence that tells what two things are being compared.

Give examples of how the two things are alike.

Add interesting details to support the examples.

Conclude by restating your main point.

1. Which sentence introduces the topic? Underline it.
2. How are Janey and Jessie alike? Find examples. Put boxes around them. Possible responses given.
3. Circle *both, alike,* and *similar* in the Student Model. Possible responses given.
4. Write the following sentence as two shorter sentences.

   Janey likes to read, but Jessie writes stories and poems.

   Possible response: Janey likes to read. Jessie writes stories and poems.

Writer's Grammar
Janey is a subject. *Jessie* is a subject. When you combine sentences, the subject and the verb must agree. Remember to change the verb when you combine subjects: (Janey *is* my best friend. Janey and Jessie *are* my best friends.).

101 Writer's Companion ■ UNIT 4 Lesson 1 *Combining Sentences*

WRITER'S GRAMMAR: SUBJECT-VERB AGREEMENT

Read aloud the Writer's Grammar. Ask a volunteer to read aloud the first sentence in the Student Model. Have students identify the subject (*Janey and Jessie*) and the verb (*are*). Point out that since the subject is plural, the verb is also plural. Remind students that a singular subject takes a singular verb. Ask students to identify the subjects and verbs. Then write the sentence *He plays ball* on the board. Ask students if the subject and verb are singular or plural (*singular*). Point out that the -*s* on *plays* is part of the *singular* form of the verb.

Evaluate a Paragraph that Compares

OBJECTIVES

- To understand how to evaluate a paragraph that compares
- To evaluate a paragraph that compares using a checklist
- To write an evaluation of a paragraph that compares

Teach/Model Discuss why it is important to evaluate writing. Explain that evaluating helps writers see what works well and what needs improvement in their writing. Tell students that the questions at the top of the page will help them evaluate a paragraph that compares.

Guided Practice Read the directions for Part A aloud. Guide students in writing responses as they refer to the Student Model for information. Point out the checklist. Tell students that a checklist is a useful tool because it focuses on key parts of writing. Explain that this checklist includes many of the writing skills practiced in this lesson.

Independent Writing Practice After reading aloud both sets of directions, have students complete Parts B and C independently.

LESSON 1: COMBINING SENTENCES

Name ______________________

Evaluating the Student Model

Evaluate a Paragraph that Compares

When you evaluate a paragraph that compares, ask yourself these questions:

- Does the writer clearly state the two things that are being compared? (Look for a topic sentence that tells what two things are being compared.)
- Does the writer provide several examples of how the two things are alike? (Look for examples of how the things are alike.)

A. Reread the Student Model on page 101. Then follow the directions below.

1. Write two more examples of how Janey and Jessie are alike. Possible responses are given.
   They are both smart. They both wear the same kind of jeans.
2. Write a detail that tells something similar, but not exactly alike about the girls.
   Janey likes to read, but Jessie writes stories and poems.

B. Now evaluate the Student Model. Check the box next to each thing the writer has done well. If you do not think the writer did a good job with something, do not check the box.

- ☐ The topic sentence clearly states what two things are being compared.
- ☐ The writer gave several examples.
- ☐ The writer gave interesting details to support the examples.
- ☐ The writer used comparison words such as *both, alike,* and *similar.*
- ☐ The writer combined sentences to make them more interesting.
- ☐ The writer concluded by restating the main point.

C. How do you think the writer could make the paragraph better? Write your ideas below.

Possible response:The writer can include more interesting details or use combining to make the comparison clearer. The last sentence should restate the main idea.

Writer's Companion ▪ UNIT 4
Lesson 1 *Combining Sentences* 102

See the rubric on page 207 for another way to evaluate the Student Model.

SHARING AND DISCUSSING

Have students compare their evaluations of the Student Model. Ask volunteers to share what they learned from the activity. Explain that they can also evaluate their own writing and their peers' writing. Then discuss the writing process and its different steps: prewriting, writing, and evaluating. Ask students what step they think follows evaluating (revising).

Revise by Making Sentences Clearer

OBJECTIVES

- To revise by combining sentences and sentence parts
- To improve an earlier piece of writing by combining sentences

Teach/Model Tell students that *revising* means improving writing by making changes. Explain that there are many ways to revise a piece of writing such as deleting or adding words. Tell students that in this lesson, they will practice revising by combining sentences and sentences parts to make sentences clearer.

Guided Practice Read aloud the introduction and the example. Discuss how deleting the names improves the sentence because it is clearer and no longer repetitive. As a class, complete Item 1.

Independent Writing Practice Have students complete the rest of Part A on their own. After a volunteer reads aloud the directions, have students do Part B independently. Invite volunteers to share their revisions aloud.

LESSON 1: COMBINING SENTENCES

Name ______________________

Revising the Student Model

Revise by Making Sentences Clearer

The student writer could have made the sentences clearer by combining sentences or sentence parts. Here is an example of how a sentence from the Student Model can be improved.

Example Also, Janey's name starts with a J. Jessie's name starts with a J.

Also, both names start with the letter J.

A. **Revise these sentences from the Student Model. Improve the sentences by combining subjects, predicates, or whole sentences.** Possible responses are given.

1. Janey and Jessie have straight dark hair. Janey and Jessie have bangs.

   Janey and Jessie have straight dark hair and bangs.

2. Janey plays piano. Jessie plays flute.

   Janey plays piano, but Jessie plays flute.

3. Janey and Jessie wear the same kind of jeans. They wear they same kind of sneakers.

   They wear the same kind of jeans and sneakers.

B. **Revise the paragraph you wrote on page 100. Combine sentences where necessary to make your writing clearer and easier to understand. Use another sheet of paper, if necessary.**

____

____

____

____

____

____

____

103 Writer's Companion ▪ UNIT 4
Lesson 1 *Combining Sentences*

WRITER'S STRATEGY: REREAD SOFTLY

Tell students that when evaluating writing, they may find it useful to reread softly their writing. Explain that when rereading softly, they should read and listen as if the writing belongs to someone else. Remind students to listen carefully for any repeated words or phrases, or for missing words.

Look at Sentence Variety

OBJECTIVES

- To understand the term *sentence variety*
- To identify sentence variety in literature

Teach/Model Read aloud the introduction and explain that writing is more interesting when sentences are varied. Tell students that writers use short and long sentences, as well as different kinds of sentences. Review three kinds of sentences and their end punctuation—statements (.), exclamations (!), and questions (?). Read aloud the passage from *Coyote Places the Stars*. As students listen, ask them to pay attention to how the writer uses statements, exclamations, questions, and different sentence lengths to make the writing interesting.

Guided Practice Read the directions for Part B aloud. Guide students in finding a statement that tells about Coyote. Then tell students to underline the statement.

Independent Writing Practice Have students complete the rest of Part B and all of Part C independently.

Use a Literature Model

LESSON 2: SENTENCE VARIETY

Name ______________________

Writer's Craft in Literature

Look at Sentence Variety

When your sentences are varied, your writing becomes more interesting to read. One way to **vary sentences is by length**. Another way is by using **different kinds of sentences**. Three kinds of sentences and their end marks are: **statements** (.), **exclamations** (!), and **questions** (?).

A. Read the Literature Model. Notice how the writer uses statements, exclamations, and questions.

Literature Model

One summer night, as he was relaxing in the cool grass with his friend Bear, Coyote had an idea. "I think I will climb to the heavens and discover their secrets!"

Bear scratched his big head and asked, "How can you do that?"

"I can get up there with no trouble at all," Coyote said.

—from *Coyote Places the Stars*
retold by Harriet Peck Taylor

B. Identify the different kinds of sentences. Possible response given.

1. Underline a statement.
2. Draw a box around an exclamation.
3. Circle a question that either Bear or Coyote asks.

C. Rewrite a sentence from the Literature Model as another kind of sentence.

Check students' work to make sure they change the kind of sentence.

____

____

Writer's Companion ■ UNIT 4
Lesson 2 *Sentence Variety* 104

SHARING AND DISCUSSING

Invite students to read their answers to Part C aloud. After each sentence is read aloud, ask volunteers to identify the sentence type (statement, question, or exclamation). Discuss how punctuation changes the meaning of the sentence.

Explore Sentence Variety

OBJECTIVES

- To deepen students' understanding of how to vary sentences by length and type
- To vary sentences by length and type

Teach/Model Use the introduction and the diagram at the top of the page to demonstrate how changing a sentence's end mark, changes its meaning. Read each sentence aloud with proper expression in order to show the meaning of each sentence.

Guided Practice Read the directions for Part A aloud and together complete the first item. Write *I have a toy poodle* on the board and ask students for suggestions on how to rewrite it. Remind students that using a variety of sentence types and lengths makes writing interesting. Repeat for Item 2.

Independent Writing Practice Have students complete Parts B and C independently.

LESSON 2: SENTENCE VARIETY

Name ______________________

A Closer Look at Writer's Craft

Explore Sentence Variety

When you change an end mark, the meaning of your sentence can change, too. Say each sentence below to yourself. Notice the extra expression added to the bolded word or words in each sentence.

Statement	Question	Exclamation
I did that.	I did **that**?	I **did** that!

A. Rewrite each sentence to vary it. Change words and end mark. Possible responses are given.

1. I have a toy poodle.
   Do you have a toy poodle?
2. I have a very nice dog.
   My dog is the best dog ever!

B. Read this sentence from *Coyote Places the Stars*. Notice the sentence type.

"I think I will climb to the heavens and discover their secrets!"

C. What would you say to coyote about this plan? Possible responses are given.

1. Write a question.
   Where will you find a ladder that tall?
2. Write a statement.
   Let me know what you learned.
3. Write an exclamation.
   That is a good idea!

105 **Writer's Companion** • UNIT 4
Lesson 2 *Sentence Variety*

EXTENDING THE CONCEPT: USING SENTENCE VARIETY

As an alternate activity for Part C, pairs of students may want to write dialogue in response to Part B. Ask partners to write five or six sentences of dialogue. Tell them to include the three kinds of sentences (statements, exclamations, and questions) and to vary sentence lengths. Partners can present their dialogue to the class.

Use Sentence Variety

OBJECTIVES

- To prepare to write a paragraph that contrasts by choosing a topic and writing contrasting details
- To practice writing different kinds of sentences with contrasting details

Teach/Model Tell students that before writing a paragraph that contrasts, they should think about two things or people they want to contrast. Read the introduction and example aloud. Suggest students use the question, *How are they different?*, as a guide when writing their sentences.

Guided Practice Read aloud the directions for Part A. Have students write the names of two things or people. Guide students to think about something that is surprising and informative about their subjects. Tell them to use this information to fill in the blanks in Part A.

Independent Writing Practice After a volunteer reads aloud the directions, have students complete Part B independently.

LESSON 2: SENTENCE VARIETY

Name ______________________

Practice with Writer's Craft

Use Sentence Variety

Before you write a paragraph that contrasts, think about how your subjects are different. You should also plan to vary your sentences when you write. Here is how one student started to plan a paragraph that contrasts his two grandmothers.

Example **Question about** *My Grandmothers*: How are they different?

Exclamation about *Grandma Bernice: She has a horse!*

Exclamation about *Grandma Melanie: Her hair is tomato red!*

Statement about *Grandma Bernice: Grandma Bernice is a vet, but she likes to sew, too.*

Statement about *Grandma Melanie: Grandma Melanie is tall and thin.*

A. **Think about two things or two people who are different from each other. Use varied kinds of sentences to plan writing about them.**

Question about ______________ **: How are they different?**

Exclamation about ______________ : ______________
Name 1

Exclamation about ______________ : ______________
Name 2

Statement about ______________ : ______________
Name 1

____

Statement about ______________ : ______________
Name 2

____

B. **Use your sentences to write a short paragraph about how the two things or people are different. Write your paragraph on another sheet of paper.**

Writer's Companion ▪ UNIT 4
Lesson 2 *Sentence Variety* 106

Reaching All Learners

Below Level · On-Level · Advanced · ELL

BELOW LEVEL

Have students work with a partner or an aide to describe contrasting details about the two things or people they chose. Then the partner or aide should help the student decide which details to write in the blanks.

ADVANCED

Encourage students to use a thesaurus to select more specific words to describe the contrasting characteristics of their subjects.

ENGLISH LANGUAGE LEARNERS

Pair less fluent English speakers with more fluent English speakers. Have students discuss contrasting things about their subjects. Have them help each other find appropriate English words and phrases to use as they fill in the blanks.

The Parts of a Paragraph that Contrasts

OBJECTIVES

- To understand how a paragraph that contrasts is organized
- To analyze a Student Model

Teach/Model Explain that the model is a first draft of a paragraph that contrasts written by a third grader. Read aloud the introduction and the Student Model. As students listen, ask them to determine if the details provide clear examples of the differences between the two grandmas.

Guided Practice Point out the items below the Student Model. Read the first item aloud and guide students in finding and underlining the topic sentence. Remind students to read the title for a clue in order to help them identify the topic.

Independent Writing Practice Have students complete Items 2 and 3 independently. Invite volunteers to share their responses with the class.

Use a Student Model

LESSON 2: SENTENCE VARIETY

Name ______________________

Focus on the Writing Form

The Parts of a Paragraph that Contrasts

A **paragraph that contrasts** tells how things or people are different. It begins with a topic sentence that clearly states what is being contrasted. Below is a draft by a third grader. As you read, think about how the student organized the paragraph.

Student Model

DRAFT

Different Grandmas
by Ethan

I have two grandmas. They are very different. One is Grandma Bernice, and one is Grandma Melanie. Grandma Melanie is tall and thin. Grandma Bernice is short and round. Grandma Melanie has long hair in a braid. Her hair is tomato red! Grandma Bernice's hair is gray and curly. Another thing that is different is where they live. Grandma Melanie lives in the city. She lives nearby in a big building. Grandma Bernice lives far away. She lives on a farm. They have different jobs, too. Grandma Melanie is a teacher. She teaches reading and writing. Grandma Bernice is a vet, but she likes to sew, too. Do you know what else? She has a horse! The horse is named Horace. My two grandmas are different. I love them the same.

Introduce two subjects to contrast.

Organize your ideas. Provide three examples for each thing you contrast.

Develop ideas by adding details about color, shape, and size.

Conclude by restating your main point.

1. Which sentences introduce the topic? Underline them.
2. Write three details, contrasting the grandmothers. Possible responses: Grandma Melanie is tall and thin. Grandma Bernice is short and round.; Grandma Melanie is a teacher. Grandma Bernice is a vet.; Grandma Melanie lives in the city. Grandma Bernice lives on a farm.
3. Circle the concluding sentences.

Writer's Grammar
Many adjectives appeal to the senses: *sight, sound, touch, taste,* and *smell.* In the sentence, *Grandma Melanie is tall and thin,* the words *tall* and *thin* are adjectives that appeal to the sense of sight. Find other adjectives in the Student Model.

107 Writer's Companion ■ UNIT 4 Lesson 2 *Sentence Variety*

WRITER'S GRAMMAR: ADJECTIVES

Read aloud the Writer's Grammar. Write the names of the five senses on the board: *sight, sound, touch, taste,* and *smell.* Then have students look for adjectives related to each sense in the Student Model. When they find an adjective, have them say which sense it belongs to. Write the word under the appropriate sense. Students can add their own examples as well. Explain that using adjectives that appeal to the senses makes writing more lively and interesting.

Evaluate a Paragraph that Contrasts

OBJECTIVES

- To understand how to evaluate a paragraph that contrasts
- To evaluate a paragraph that contrasts using a checklist
- To write an evaluation of a paragraph that contrasts

Teach/Model Discuss the benefits of evaluating writing. Explain that evaluating writing identifies ways that the writing is effective and also shows its weaknesses. Use the questions at the top of the page to guide students as they evaluate a paragraph that contrasts.

Guided Practice Read aloud the directions for Part A. Guide students to refer to the Student Model for information in order to help them answer the questions. Point out the checklist. Explain that a checklist will help them evaluate writing because it focuses on key points in the paragraph.

Independent Writing Practice After reading aloud both sets of directions, have students complete Parts B and C independently.

LESSON 2: SENTENCE VARIETY

Name ______________________

Evaluating the Student Model

Evaluate a Paragraph that Contrasts

When you evaluate a paragraph that contrasts, ask yourself these questions:

- Does the writer introduce two subjects to contrast? (Look for a topic sentence.)
- Does the writer provide three examples? (Look for sentences that tell how the subjects are different.)
- Does the writer use varied sentence types? (Look for different end marks and conjunctions.)

A. Reread the Student Model on page 107. Then answer these questions. Possible responses

1. Write two examples that contrast the way the grandmas look.

   different hair, different size

2. Write two details that appeal to the senses.

   hair is tomato red, lives nearby in a big building

B. Now evaluate the Student Model. Check the box next to each thing the writer has done well. If you do not think the writer did a good job with something, do not check the box.

- ☐ The writer included a topic sentence that tells what is being contrasted.
- ☐ The writer described differences using three examples.
- ☐ The writer gave interesting details to support the examples.
- ☐ The writer used different kinds of sentences.
- ☐ The writer concluded by restating the main point.

C. How do you think the writer could make the paragraph of contrast better? Write your ideas below.

Possible response: The writer can use more sentence variety. The writer can give more details.

Writer's Companion ■ UNIT 4
Lesson 2 *Sentence Variety*
108

See the rubric on page 207 for another way to evaluate the Student Model.

SHARING AND DISCUSSING

Have students compare their answers from Part A. Discuss if the writer gave a clear picture of each of the grandmas. Discuss if the details and examples were easy to understand. Ask students for examples of other kinds of details that could be included.

Revise by Making Contrasts Clearer

OBJECTIVES

- To revise a paragraph that contrasts by making contrasts clearer
- To revise by combining sentences and varying sentence lengths or types

Teach/Model Tell students that writers revise their work to make their sentences clear and to be sure the sentences make sense. Tell students that in this lesson, they will revise by making contrasts clearer through combining sentences.

Guided Practice Read aloud the introduction and the example. Have students return to the model on page 107. Guide them in recognizing that when the sentences are combined, there is less repetition and the sentences make more sense. Guide students as they revise Items 1–2 in Part A.

Independent Writing Practice Have students complete Part B independently. Invite volunteers to share their revisions aloud.

LESSON 2: SENTENCE VARIETY

Name ______________________

Revising the Student Model

Revise by Making Contrasts Clearer

The student writer should make the contrasts clearer and easier to understand. One way to do this is by combining sentences. Here is an example of how a sentence from the Student Model can be improved.

Example I have two grandmas. They are very different.

I have two grandmas who are very different.

A. **Read each pair of sentences from the Student Model. Combine the pair into one sentence.** Possible responses are given.

1. Grandma Melanie is tall and thin. Grandma Bernice is short and round.

   Grandma Melanie is tall and thin, but Grandma Bernice is short and round.

2. My two grandmas are different. I love them the same.

   My two grandmas are different, but I love them the same.

B. **Revise the paragraph you wrote on page 106. Combine and vary your sentences. Use another sheet of paper, if necessary.**

Answers will vary. Students should use various sentence types and words that show contrasts.

109

Writer's Companion ■ UNIT 4
Lesson 2 *Sentence Variety*

WRITER'S STRATEGY: CONJUNCTIONS

Remind students that using conjunctions is one way to combine short sentences into longer ones. Have students read aloud their revised sentences from Part A. Discuss which conjunction was used. Replace that conjunction with another conjunction. Discuss with students if this changes the meaning of the sentence. Remind students that when they rewrite, they should consider each word and make changes and revisions that will make the meaning clear.

Look at More About Sentence Variety

OBJECTIVES

- To understand what is meant by *varying sentence length* and *sentence variety*
- To identify sentence variety in literature

Teach/Model Read aloud the introduction. Explain that writers vary the length of their sentences and use different kinds of sentences to make their writing interesting. Explain that sentences may be short or long and that different kinds of sentences include statements, exclamations, and questions. Read aloud the passage from *Why Mosquitoes Buzz in People's Ears*. As students listen, ask them to notice how the writer uses a variety of sentences.

Guided Practice Read the directions for Part B aloud. Guide students in identifying the different kinds of sentences used in the passage. First, ask students to find and underline the conjunctions (Item 1). Then guide students in finding an exclamation (Item 2) and a question (Item 3).

Independent Writing Practice After reading aloud the directions, have students complete Part C independently.

Use a Literature Model

LESSON 3: MORE ABOUT SENTENCE VARIETY

Name ______________________

Writer's Craft in Literature

Look at More About Sentence Variety

Writers use a mix of long sentences and short sentences. They use statements, exclamations, and questions. **Sentence variety** makes writing more interesting.

A. Read the model. Notice how the writer uses a variety of sentences.

Literature Model

The big snake raised his head and said, "Good morning, Iguana."

The iguana did not answer but limbered on, bobbing his head, badamin, badamin.

"Now why won't he speak to me?" said the python to himself. "Iguana must be angry about something. I'm afraid he is plotting some mischief against me!"

—from *Why Mosquitoes Buzz in People's Ears*
retold by Verna Aardema

B. Identify different kinds of sentences.

1. Underline the conjunctions.
2. Draw a box around an exclamation.
3. Circle a question the python asks.

C. Which two sentences can be combined into one? Combine them below. Use proper punctuation and end marks.

Possible response: "I'm afraid Iguana is angry about something and is plotting some mischief against me!

Writer's Companion ■ UNIT 4
Lesson 3 *More About Sentence Variety* 110

SHARING AND DISCUSSING

Invite students to read their answers to Part C aloud. Have them share what kind of sentence they wrote. Ask if the sentence is a statement, a question, or an exclamation. Discuss what punctuation was used.

Explore More About Sentence Variety

OBJECTIVES

- To identify sentence types and end marks
- To identify conjunctions in a long sentence
- To rewrite a long sentence into several short sentences

Teach/Model Use the diagram at the top of the page to explain sentence variety (statements, questions, and exclamations) and sentence length. Tell students that using different sentence lengths and types makes writing more interesting.

Guided Practice Read the directions for Part A aloud. Model answering Item 1. Read aloud the sentence three times, but each time read it with different end punctuation. Guide students in identifying the type of sentence. Remind students that changing a sentence's end mark will change its meaning. Complete Items 2–5 together with students.

Independent Writing Practice After a volunteer reads aloud both sets of directions, have students complete Parts B and C independently.

LESSON 3: MORE ABOUT SENTENCE VARIETY

Name ______________________

A Closer Look at Writer's Craft

Explore More About Sentence Variety

Sentence variety makes writing more interesting.

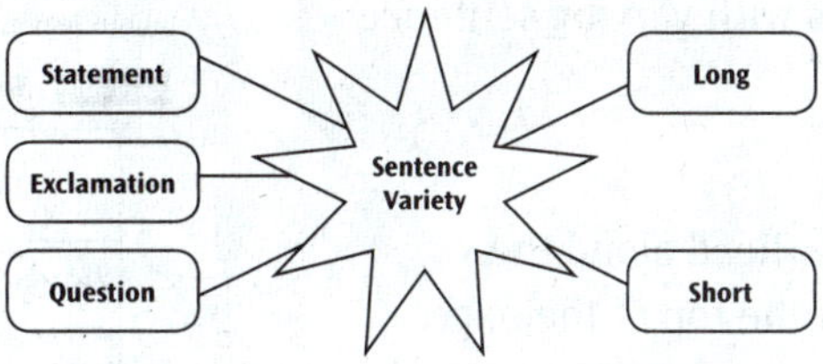

A. Draw a line from the sentence to the word that describes it. Then write the correct end mark on the line.

combined	1. The corn grew very tall last summer .
exclamation	2. Do you know my sister ?
question	3. Our friend is moving away, and we will miss him .
short	4. Your hat is huge !
statement	5. Squirrels scampered .

B. Read the combined sentence from *The Ant and the Grasshopper.* Underline the conjunction. On the lines, rewrite the sentence as two or three separate sentences.

Winter arrived a week later <u>and</u> brought whirls of snow and ice.

1. Winter arrived a week later.
2. Winter brought whirls of snow.
3. Winter brought ice.

C. Rewrite one of the short sentences about winter as a different kind of sentence.

Possible response: Winter brought so much ice!

EXTENDING THE CONCEPT: USING SENTENCE VARIETY

After completing Parts B and C, have groups of five students write their sentences on the board. As a class, work to combine, shorten, and change the sentence type within each group of sentences. Use conjunctions as necessary. Review the revised sentences with the class. Then compare the revised sentences to the original sentences in Part B.

Use More About Sentence Variety

OBJECTIVES

- To prepare to write a paragraph that explains by selecting a topic and generating sentences
- To practice writing different kinds of sentences with varying sentence length

Teach/Model Read aloud the information at the top of the page. Invite volunteers to read the example. Tell students that they should plan to write a paragraph that explains by choosing a topic and writing details. Explain that students can then write different kinds of sentences with varying length.

Guided Practice Read aloud the directions for Part A. Remind students to choose something they already know how to explain for their topic. Guide them in writing each kind of sentence. Tell students to use correct end punctuation.

Independent Writing Practice Have students complete Part B independently. Ask volunteers to share their paragraphs.

LESSON 3: MORE ABOUT SENTENCE VARIETY

Name ______________________

Practice with Writer's Craft

Use More About Sentence Variety

Before you write a paragraph that explains, think about the topic and details you want to include. Here is how one student planned a paragraph about leaves.

Example **Topic:** *Leaves*

Question	Statement	Exclamation	Combined Sentence
Have you ever noticed that some leaves change color at different parts of the year?	*This happens every year.*	*This is how trees get ready for winter!*	*The leaves can be yellow, red, or purple.*

A. Think about something you would like to explain. Write your main idea. Then fill out the rest of the chart.

Topic: ______________________

Question	Statement	Exclamation	Combined Sentence

B. Use information from your chart to write a short paragraph that explains. Use another sheet of paper. Answers will vary, but should show sentence variety.

Writer's Companion • UNIT 4
Lesson 3 *More About Sentence Variety* 112

Reaching All Learners

BELOW LEVEL

Have students work with a partner or an aide to state their topic and main idea. Then work together to write a question, a statement, and an exclamation about their topic. Discuss end marks. Help students fill out their charts.

ADVANCED

Encourage students to write two or three sentences of each type. Remind them to vary sentence length and to verify that they are using correct punctuation. Then have students trade with a partner and review each other's work.

ENGLISH LANGUAGE LEARNERS

Pair less fluent English speakers with more fluent English speakers. Have students brainstorm ideas for writing a question, a statement, and an exclamation about their topic before filling in their charts.

The Parts of a Paragraph that Explains

OBJECTIVES

- To understand how a paragraph that explains is organized
- To analyze a Student Model

Teach/Model Explain that the model on this page is a first draft of a paragraph that explains written by a third grader. Read aloud the introduction and the steps that explain how to organize this type of paragraph. Emphasize that a paragraph that explains begins with a sentence that clearly introduces the topic.

Guided Practice Point out the items below the Student Model. Read aloud the first item and guide students in explaining the writer's purpose for writing *Leaves*.

Independent Writing Practice Have students complete Items 2–5 independently. Invite volunteers to share their responses.

Use a Student Model

LESSON 3: MORE ABOUT SENTENCE VARIETY

Name ______________________

Focus on the Writing Form

The Parts of a Paragraph that Explains

The purpose of a **paragraph that explains** is to tell about something. The paragraph starts with a topic sentence that gives the main idea. The supporting sentences give details and information. Below is a draft by a third grader. As you read, think about the parts of the paragraph and the varied sentences.

Student Model

DRAFT

Leaves

by Jenny Sue

Have you ever noticed that some leaves change color at different parts of the year? This is how trees get ready for winter! Sometimes the leaves look green. At the end of summer, some trees stop feeding their leaves. This is how the other colors come out. The leaves can be yellow or red. The leaves can be purple, too. Leaves turn red if they have a lot of sugar in them. Then, some trees lose their leaves. This happens every year.

Begin with a sentence that clearly introduces your topic.

Add facts, examples, and details.

Put your ideas in an **order** that makes sense.

Conclude by restating your main idea.

1. What is the writer's purpose for writing this paragraph?

   to explain why leaves change colors

2. Which sentence introduces the topic? Underline it.
3. Double underline a fact and detail that supports the fact.
4. Which two short sentences could be combined into a longer one? Combine them below.

   Possible response: The leaves can be yellow, red, or purple.

5. Put a box around the concluding sentence.

Writer's Grammar
Too and *also* have the same meaning. Put a comma (,) before *too* when it's the last word in the sentence. Find *too* in the Student Model.

113 Writer's Companion • UNIT 4
Lesson 3 *More About Sentence Variety*

WRITER'S GRAMMAR: *TOO* AND *ALSO*

Read aloud the Writer's Grammar. Have a volunteer find *too* in the Student Model. Point out that a comma (,) comes before the word. Have students rewrite the sentence *The leaves can be purple, too.*

Tell students to replace the word *too* with the word *also*. Ask them where they would put *also* in the sentence. Would they put it in the same place as *too*, or would it sound better elsewhere?

Evaluate a Paragraph that Explains

OBJECTIVES

- To understand how to evaluate a paragraph that explains
- To evaluate a paragraph that explains using a checklist
- To write an evaluation of a paragraph that explains

Teach/Model Discuss why it is important to evaluate writing. Explain that evaluating helps writers determine whether all the steps are in an order that makes sense. Use the points at the top of the page to discuss what to look for when evaluating a paragraph that explains.

Guided Practice Read aloud the directions for Part A. Guide students in writing answers as they refer to the Student Model for information. Point out the checklist in Part B. Explain that the checklist will help students look at each part of the paragraph and evaluate each part.

Independent Writing Practice After reading aloud both sets of directions, have children complete Parts B and C independently.

LESSON 3: MORE ABOUT SENTENCE VARIETY

Name ________________________

Evaluating the Student Model

Evaluate a Paragraph that Explains

When you evaluate a paragraph that explains, ask yourself these questions:

- Does the writer clearly state the main idea?
  (Look for a topic sentence that tells what is being explained.)
- Does the writer provide facts and details that explain the main idea?
  (Look for examples and interesting details.)

Possible responses are given.

A. Reread the Student Model on page 113. Then answer these questions.

1. Write a sentence that tells something about leaves. Leaves turn red if they have a lot of sugar in them.
2. Write two details that support the idea that leaves change color to get ready for winter. At the end of summer, some trees stop feeding their leaves. This is how the colors come out. Then, some trees lose their leaves.

B. Evaluate the Student Model. Put a check in the box next to each thing the writer has done well. If you do not think the writer did a good job, do not check the box.

- ☐ The writer included a topic sentence that clearly states what is being explained.
- ☐ The writer provided facts and examples to support the main idea.
- ☐ The writer used different kinds of sentences.
- ☐ The writer concluded by restating the main idea.

C. How do you think the writer could make the paragraph better? Write your ideas below.

Possible response: The writer could have included more details.

____

____

____

Writer's Companion • UNIT 4
Lesson 3 *More About Sentence Variety* 114

See the rubric on page 207 for another way to evaluate the Student Model.

SHARING AND DISCUSSING

Discuss with students the benefits of evaluating writing. Explain that when they evaluate someone else's writing, it may help them improve their writing as well. Have students share their experiences. Ask if it was helpful to see comments and suggestions. Have students tell which types of comments are the most useful.

Revise by Adding Details

OBJECTIVES

- To revise sentences by adding details
- To revise by combining sentences and varying sentence lengths

Teach/Model Tell students that when they revise, they should think about how to make their writing clearer and more detailed. Explain that in this lesson they will add details to make sentences clearer.

Guided Practice Read aloud the introduction and the example. Have students return to the Student Model on page 113. Ask them why it is important to revise the sentence *Sometimes the leaves look green* by adding *In the spring and summer*. Guide students to understand that the addition of details makes the sentence clearer. Have students add words from the Word Bank to the sentence in Part A, Item 1. Remind them that their responses should be clearer and more detailed.

Independent Writing Practice Have students complete the rest of Part A on their own. After reading aloud the directions, have students do Part B independently.

LESSON 3: MORE ABOUT SENTENCE VARIETY

Name ____________________

Revising the Student Model

Revise by Adding Details

One way the writer can make the paragraph more interesting is by adding details. Here is how a sentence from the Student Model can be improved.

Example Sometimes the leaves look green.

In the spring and summer, the leaves look green.

A. **Revise these sentences from the Student Model. Make the sentences clearer and more detailed. Use words from the Word Bank.** Possible responses are given.

1. This is how the other colors come out.

   This is how the other colors that are hidden underneath come out.

2. Then, some trees lose their leaves.

   In the winter, some trees lose their leaves.

3. This happens every year.

   This happens every year, so that the plant can be new and fresh for springtime.

Word Bank
fresh
hidden underneath
in the winter
new
pretty
springtime

B. **Revise the paragraph you wrote on page 112. Add some interesting details. Remember to vary and combine your sentences. Use another sheet of paper, if you need more space.**

Answers will vary. Students should add interesting details, and use different sentence types.

115 Writer's Companion ▪ UNIT 4
Lesson 3 *More About Sentence Variety*

WRITER'S STRATEGY: FACTS, EXAMPLES, AND DETAILS

Remind students that when they write a paragraph that explains, they need to include facts, details, and examples. Explain that a *fact* is true. Tell them that an *example* is a sample or a model to be imitated. Often the words, "For example…" are used before the example is given. Explain that a *detail* can be something that appeals to the senses, or that describes a fact. Tell students to add facts, examples, and details to improve their writing.

Review Sentence Fluency

OBJECTIVES

- To review sentence variety including types and lengths of sentences and sentences combining
- To identify statements, exclamations, questions, and short and longer combined sentences in literature

Teach/Model In this lesson, students will review what they have learned about sentence fluency including types and lengths of sentence and combining sentences. Read aloud the introduction and the directions for Part A. Then read aloud the passage from *Lon Po Po*. As students listen, ask them to notice the use of sentence variety in the passage.

Guided Practice Read the directions for Part B aloud. Guide students to identify and underline a statement.

Independent Writing Practice Have students complete the rest of Part B on their own. After reading aloud the directions, have students do Part C independently.

Use a Literature Model

LESSON 4: REVIEW SENTENCE FLUENCY

Name ______________________

Writer's Craft in Literature

Review Sentence Fluency

Good writers vary their sentences to make their writing interesting. They use statements, exclamations, and questions. They use short sentences and longer, combined sentences.

A. Below are sentences from a story about three sisters who outsmart a wolf. Notice the sentence variety as you read.

Literature Model

But an old wolf lived nearby and saw the good mother leave. At dusk, disguised as an old woman, he came up to the house of the children and knocked on the door twice: bang, bang.

Shang, who was the eldest, said through the latched door, "Who is it?"

"My little jewels," said the wolf, "this is your grandmother, your Po Po."

"Po Po!" Shang said. "Our mother has gone to visit you!"

—from *Lon Po Po*
retold by Ed Young

B. Review sentence variety. Possible responses given.

1. Underline a statement.
2. Put a box around a question asked by a character.
3. Circle an exclamation.

C. Choose a longer, combined sentence from the model. Write the two short sentences that the longer sentence might have come from. Answers will vary.

But an old wolf lived nearby. The old wolf saw the good mother leave.

____

____

Writer's Companion ■ UNIT 4
Lesson 4 *Review Sentence Fluency* 116

SHARING AND DISCUSSING

Have volunteers share their responses to Part C aloud. Discuss why the original long sentence may have been used. Explain that skilled writers use sentence variety to avoid repetition and to make their writing lively.

Review Sentence Fluency

OBJECTIVES

- To review sentence variety including types and lengths of sentences and sentence combining
- To add words and variety to make sentences more interesting

Teach/Model Remind students that sentence variety means using different kinds of sentences. Point out that this includes short or long, combined sentences, statements, questions, and exclamations. Use the graphic organizer to review sentence variety with students. Read aloud the directions for Part A and the sentence from *Lon Po Po*. As students listen, ask them to notice the length and type of sentence. Then read aloud the directions for Part B. Have students identify clear ideas, so they can rewrite the long sentence into two short sentences. Guide students as they complete Part B.

Guided Practice Read the directions for Part C aloud. Suggest that students add vivid words and details to rewrite this sentence.

Independent Writing Practice Have students complete the rest of Part C independently.

LESSON 4: REVIEW SENTENCE FLUENCY

Name ______________________

A Closer Look at Writer's Craft

Review Sentence Fluency

In this unit, you learned about different kinds of sentences.

Sentence Variety means using different kinds of sentences.

Short Sentences | **Combined Sentences** | **Statements** | **Questions** | **Exclamations**

A. Read this sentence from *Lon Po Po.*

The journey is long, my children, and the day is short.

B. Write the combined sentence as two short sentences. Possible responses given.

1. The journey is long, my children.
2. The day is short.

C. Rewrite each of these sentences from *Lon Po Po.* Add words and variety to make the sentences more interesting. Possible responses given.

1. The wolf was furious.

   The big, bad wolf was furious that he could not enter.

2. Have you eaten gingko nuts?

   You must have eaten ginko nuts!

3. Soon the old wolf pretended to be sleepy.

   Is the wolf about to fall asleep?

117 Writer's Companion ■ UNIT 4
Lesson 4 *Review Sentence Fluency*

EXTENDING THE CONCEPT: ADDING DETAILS AND VIVID WORDS

Before students complete Part C, have them recall the story of *Lon Po Po*. Ask students to imagine what the wolf looked like, and what he was feeling when he was furious. Have them brainstorm other words for *furious*. Write the words on the board. Then have students brainstorm vivid verbs that could replace *was* in Sentence 1 in Part C. Challenge them to use these words to rewrite the sentence. Continue in the same manner for Sentences 2 and 3. Have students read their rewritten sentences aloud. Discuss how adding details and vivid words makes sentences more interesting.

Review Sentence Fluency

OBJECTIVES

- To identify a topic and supporting details
- To prepare to write a compare and contrast essay by using a Venn diagram
- To write sentences that compare and contrast

Teach/Model Read aloud the introduction. Tell students that before they write a compare and contrast essay, they will learn how to use a Venn diagram. Use the example to teach the concept. Point out that the outer part of each circle shows the differences, or contrasts. The center shows how the stories are alike, or compares. Have students suggest other points to add to the diagram.

Guided Practice Read aloud the directions for Part A. Guide students to identify the characters in both stories and to compare the wolves.

Independent Writing Practice Have students complete Part B independently.

LESSON 4: REVIEW SENTENCE FLUENCY

Name ______________________

Practice with Writer's Craft

Review Sentence Fluency

Before you write an essay that compares and contrasts, think about what your topic and supporting details will be. Then think about the different kinds of sentences you will use to add interest to your writing. Here is how one third grader planned an essay explaining how two folktales are alike and different.

Example

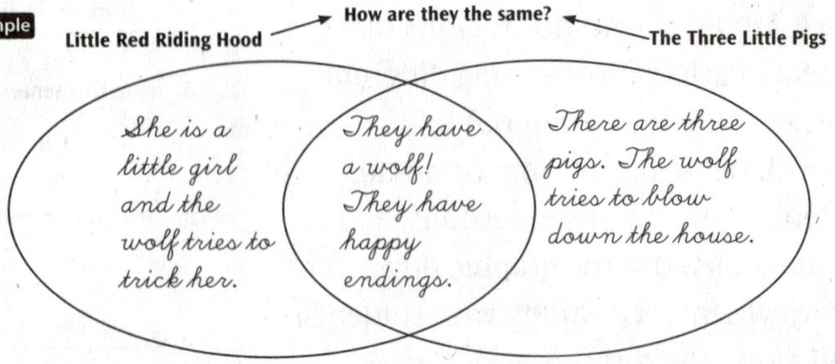

A. Answer the questions based on the writing plan.

1. Who are the main characters in *Little Red Riding Hood*? Little Red Riding Hood, the wolf
2. Who are the main characters in *The Three Little Pigs*? the three little pigs, the wolf
3. How are the wolves alike in both stories? They both want to cause trouble.

B. Write two sentences—one that compares and one that contrasts something from both stories.

Possible response: Both stories are about characters against a wolf. Little Red Riding Hood is a human girl, and the the three little pigs are animals.

Writer's Companion • UNIT 4
Lesson 4 *Review Sentence Fluency* 118

Reaching All Learners

BELOW LEVEL

Help students identify things that are similar and things that are different in the stories. Then have students complete these sentence frames: Both stories are about ________________. Little Red Riding Hood is ______, and the three little pigs are _____.

ADVANCED

Encourage students to write additional sentences that compare and contrast things from the stories. Suggest they use sentence variety in their writing. Remind them to use correct punctuation for each kind of sentence.

ENGLISH LANGUAGE LEARNERS

Pair less fluent English speakers with more fluent English speakers. Have students discuss things that are similar and things that are different in the stories. Have them help each other find appropriate English words to use in their sentences.

The Parts of a Compare and Contrast Essay

OBJECTIVES

- To understand how a compare and contrast essay is organized
- To analyze a Student Model

Teach/Model Explain to students that the writing model on this page is a first draft of a third grader's compare and contrast essay. Read aloud the steps for how to organize a compare and contrast essay. Explain that the introduction states the topic and the main idea of the essay. Emphasize that one main idea states similarities and one main idea states differences. Remind students that both main ideas need to be supported with examples. Then read the model aloud. Have children listen for the supporting examples.

Guided Practice Point out the items below the Student Model. Read aloud the first item. Guide students to write a sentence that tells the topic of the essay.

Independent Writing Practice Have students complete Items 2–4 independently. Then have them share their responses and discuss the meaning of *alike* and *different*.

How to Organize a Compare and Contrast Essay

Name ____________________

Focus on the Writing Form

The Parts of a Compare and Contrast Essay

A **compare and contrast essay** tells how two things are alike and how they are different. Below is a draft written by a third grader. As you read, think about how the student organized the essay.

Student Model

DRAFT

Two Stories
by Leticia

Little Red Riding Hood and *The Three Little Pigs* are two stories that are alike and different. Do you want to know how they are similar? [Both have a wolf in it! The wolf is bad and scares them. The wolf loses in the end and both stories have happy endings.]

Here is how the stories are different. Red Riding Hood is a little girl. *The Three Little Pigs* has three pigs. The wolf tries to trick Red Riding Hood, but the wolf tries to blow down the houses of the three pigs. So that's how these two stories are alike and different.

Introduce the topic and state the main idea of the essay.

State a **main idea** of how the two things are **alike**. Give an example to support that idea.

State a **main idea** of how the two things are **different**. Give an example to support that idea.

Conclude by restating your main idea.

1. What is the topic of the essay?
   The two stories are alike and different.
2. Find sentence that give examples of how the stories are **alike**. Put [ ] around it.
3. Find sentences that give examples of how the stories are **different**. Draw a box around them. Possible response given.
4. Circle the concluding sentence.

Writer's Grammar
Remember that the article *a* changes to *an* when the noun it describes starts with a vowel. For example, "I have *a* banana, but not *an* orange." Find the use of *a* in the model.

WRITER'S GRAMMAR: *A* AND *AN*

Write this sentence on the board: *I had to wait for ____ hour.* Ask students what word goes in the blank *(an)*. Remind them that they have learned that the article *a* changes to *an* when the noun it describes starts with a vowel. Point out that *hour* does not start with a vowel, but because the *h* is not pronounced, the word starts with the vowel sound *ou*. Explain that there are often exceptions to rules in English and most languages.

Evaluate a Compare and Contrast Essay

OBJECTIVES

- To understand what to look for when evaluating a compare and contrast essay
- To see how a 4-point rubric is used to evaluate writing
- To find out why one Student Model received a score of 4 (excellent)

Teach/Model Explain to students that some of the writing they do at school and on writing tests will be evaluated using a *rubric.* Explain that a rubric is a chart that evaluates writing on a scale of points.

Guided Practice Read aloud the directions for Part A. Then read aloud the Student Model. Have students listen, without looking at the Student Model, and think about why this Student Model is well done. Then have students read the model and the teacher's comments that refer to it. Explain that the comments show some of the ways a teacher might evaluate a compare and contrast essay. Ask students to identify comments that have to do with sentence variety, development, or word choice.

LESSON 4: REVIEW SENTENCE FLUENCY

Name ______________________

Evaluate the Writer's Form

Evaluate a Compare and Contrast Essay

A. Two students were asked to write an essay comparing and contrasting two folktales. This essay got a score of 4. When using a 4-point rubric, a score of 4 means "excellent." Read the compare and contrast essay and the teacher comments that go with it. Find out why it is a success.

Student Model

DRAFT

Two Stories
by Seth

The Tortoise and the Hare and *The Grasshopper and the Ants* are alike. They are also different. Both stories have animals that act like people. In both, one character works hard and one does not. The one that works hard wins!

What are the differences? In *The Grasshopper and the Ants,* they are getting ready for the winter, but in *The Tortoise and the Hare* they are having a race. In *The Grasshopper and the Ants,* the grasshopper wants to have fun, but in the other story the hare is playing a trick. So even though they are different, they teach the same lesson.

Great! You introduced the topic in the first sentence.

Yes! You used sentences that vary in length.

Good! You used a variety of sentences.

Way to go! You used examples and details to support your main points.

Good job! You have restated the main point at the end.

Writer's Companion ■ UNIT 4
Lesson 4 *Review Sentence Fluency* 120

WRITER'S STRATEGY: READ IT THREE TIMES

Tell students that when they evaluate their own writing, they should read it carefully and think about its organization, sentence variety, and word choice. Suggest that students reread their own work three times when they are in the revising stage of the writing process. Explain that with each reading, they will be able to focus on one specific item. Suggest to students that they might also want to read their writing aloud.

Evaluate a Compare and Contrast Essay

OBJECTIVES

- To read a Student Model that received a score of 2 *(needs improvement)*
- To understand what makes a compare and contrast essay score poorly on a rubric
- To use a rubric to evaluate a compare and contrast essay

Teach/Model Tell students they will read a compare and contrast essay that received a score of 2 *(needs improvement)* on a 4-point rubric. Read the Student Model aloud. Have students listen without looking at the writing, and think about why this essay needs improvement.

Guided Practice Discuss with students any strengths or weaknesses they noticed while listening to the essay. Then have students read the model and the teacher's comments. Invite a volunteer to summarize the comments and point out problems with variety, development, or word choice.

Independent Writing Practice Explain how to use the 4-point rubric. Refer to the Extending the Concept feature. Then have students complete the rubric independently.

Evaluate a Student Model

LESSON 4: REVIEW SENTENCE FLUENCY

Name ______________________

Evaluate the Writer's Form

B. The essay below got a score of 2. When using a 4-point rubric, a score of 2 means "needs improvement." Read the essay and the teacher comments that go with it. Find out why it received a score of 2.

Student Model

DRAFT

Two Stories
by Philip

I read two stories. They are *The Emperor's New Clothes* and *The Peacock and the Crane*. They are alike. They are different. Someone in them thinks that looking good is more important than anything, but they are wrong. There are differences, too. Like the emperor tries to trick people, but the peacock doesn't try to trick people. So they are alike and different. I liked the stories very much.

Can you introduce the topic and main idea in the first sentence?

These two sentences might be combined into a longer one.

Try putting the contrasts in another paragraph.

Nice example of contrast. Can you think of another way the stories are different?

This last sentence is an opinion. You need to restate your main idea.

C. This chart is an example of a 4-point rubric. A score of 4 is the highest score in this rubric. A score of 1 is the lowest score. Use the teacher comments above to complete the rubric. Write a number score on the line in the first column.

	4	3	2	1
Sentences ______	☐ The writer varies the sentences.	☐ The writer varies some of the sentences.	☐ The writer does not vary many of the sentences.	☐ The writer does not vary the sentences at all.
Development ______	☐ The essay has a clear main idea and strong details.	☐ The essay has a main idea and is supported by details.	☐ The essay does not have a clear main idea and has few supporting details.	☐ The main idea is not clear and there are few or no supporting details.
Word Choice ______	☐ The writer uses clear, exact words and phrases. The writing is interesting to read.	☐ The word choices are clear. The writer uses some interesting words and phrases.	☐ The writer does not use words or phrases that make the writing clear to the reader.	☐ The writer uses word choices that are unclear or poor.

EXTENDING THE CONCEPT: USING A RUBRIC TO EVALUATE

Direct students' attention to the rubric. Read aloud each box in the *Sentences* row. Have students check the box next to the information that best describes the writer's use of sentence variety. Then point out the number at the top of the column that contains the checked box. Tell students to write that number in the blank below the words *Sentences*. Repeat for *Development* and *Word Choice*.

Extended Writing/Test Prep

OBJECTIVES
- To apply the craft of sentence fluency in a longer piece of writing
- To select a topic for an extended piece of writing and begin the prewriting process
- To become familiar with the kinds of writing prompts that appear on writing assessments

Teach/Model Tell students that the next two pages will provide them with an opportunity to choose a topic for a longer piece of writing. Then they will follow the steps of the writing process to complete it. Tell students that they can choose from three possible writing activities. Discuss the choices listed in Part A and have each student select one.

Guided Practice Review the writing forms covered in the theme. Then have students reread the Student Model tips in the lesson where that writing form was taught:
- A Paragraph that Compares: Lesson 1, Page 101
- A Paragraph that Contrasts: Lesson 2, Page 107
- A Paragraph that Explains: Lesson 3, Page 113
- A Compare and Contrast Essay: Lesson 4, Page 119.

Have students begin the prewriting process as you guide them in completing Part B.

Extended Writing/Test Prep

Name ______________________________

Extended Writing/Test Prep

On the last two pages of this lesson, you will use what you have learned about sentence variety to write a longer written work.

A. Read the three choices below. Put a star by the writing activity you would like to do.

1. Respond to a Writing Prompt

   Writing Situation: Your classmate missed a day of school and asks you to tell about two stories you read in class.

   Directions for Writing: Think about two folktales you know. Now write a compare and contrast essay about how they are alike and different. Use what you have learned about sentence variety when you write your essay.

2. Choose one of the pieces of writing you started in this unit:
   - a paragraph that compares (page 100)
   - a paragraph that contrasts (page 106)
   - a paragraph that explains (page 112)

   Expand your beginning into a complete piece of writing. Use what you have learned about different kinds of sentences in your writing.

3. Choose two favorite meals or foods. Write an essay that compares and contrasts them. Use different types of sentences in your writing.

B. Use the space below and on the next page to plan your writing.

TOPIC: ______________________________

____

WRITING FORM: ______________________________

HOW WILL I ORGANIZE MY WRITING: ______________________________

____

WRITER'S STRATEGY: CHOOSING A TOPIC

To help students choose a topic, suggest that they ask themselves questions such as these: What do I know about the topic? How much would I enjoy writing about the topic? Do I have enough to say about the topic to write about it? Do I have enough time to put my ideas together to write about this topic? After answering these questions, students should brainstorm ideas for their writing.

Extended Writing/Test Prep

OBJECTIVES

- To use a graphic organizer to plan a longer piece of writing
- To use the steps of the writing process to complete a longer piece of writing independently

Teach/Model Explain to students that in Part C, they will draw a graphic organizer to help them plan their writing. Provide each student with a photocopy of the graphic organizer from the back pages of this guide. Then model how to complete it.

Guided Practice: Guide students in completing their graphic organizers. Remind them that a graphic organizer is a planning tool. Explain that they do not have to write every detail at first. They can add more details as they write.

Independent Writing Practice: **Draft, Revise, Publish** Tell students to write the draft on a separate sheet of paper. After the first draft is finished, ask them to revise their work. Guide them to use vivid words and to vary sentences. Then invite them to illustrate their writing. Compile class books for each writing form. Students can read the books during a special class writing celebration.

LESSON 4: REVIEW SENTENCE FLUENCY

Name ______________________________

Extended Writing/Test Prep

C. **In the space below, draw a graphic organizer that will help you plan your writing. Fill in the graphic organizer. Use the lines below to write your notes.**

Notes

____

____

____

____

____

____

____

____

D. **Write your essay on another sheet of paper.**

123 **Writer's Companion** ▪ UNIT 4
Lesson 4 *Review Sentence Fluency*

SHARING AND DISCUSSING

Have students discuss what they were thinking about as they completed their graphic organizers. Ask how the graphic organizers helped them prepare to write. Then have students point out specific parts of their final drafts and explain which part of the graphic organizer they used as they wrote each part.

Answering Multiple-Choice Questions

OBJECTIVES

- To understand that some writing tests contain multiple-choice questions
- To use knowledge of articles to answer questions

Teach/Model Tell students that some writing tests include multiple-choice questions. Explain that this lesson will help them develop their ability to read and answer this type of question well. Read aloud the directions for Part A. Then read the test tip. Model answering Question 1 by writing the item on the board. Identify the articles in each answer choice as you model how to determine the correct answer. Show students how to fill in the bubble next to the correct answer.

Guided Practice Read aloud Question 2. Tell students to look for the articles as they read the answer choices. Make certain they understand that only one of the choices is correct.

Independent Writing Practice Have students complete Question 2 independently. Remind them to mark the bubble next to the correct answer.

Name ______________________________

Writing Test Practice

Answering Multiple-Choice Questions

Some writing tests have questions with answer choices. This lesson will help you practice this kind of test.

A. Some test questions may ask you to put sentences in order. Read the test tip. Then read each question. Mark the circle next to the correct answer.

1. Put the ideas in the box together to create a sentence that makes sense.

there is a
my jacket
hole in
big

Which sentence correctly combines the words from the box?

- Ⓐ There is a my jacket hole in big.
- Ⓑ There is a hole in big my jacket.
- Ⓒ There is a big hole in my jacket.

2. Put the ideas in the box together to create a sentence that makes sense.

apple
brown bowl
from the
I took an

Which sentence correctly combines the words from the box?

- Ⓕ From the apple brown bowl I took an.
- Ⓖ I took an apple from the brown bowl.
- Ⓗ I took an brown bowl from the apple.

Test Tip:
The article *a* usually comes before a word that begins with a consonant (*a* bear; *a* pot). The word *an* usually comes before a word that begins with a vowel (*an* egg; *an* ox).

SHARING AND DISCUSSING

Pair students. Have them compare their responses to the last test question. Ask them to share their thinking process in selecting each answer.

Answering Multiple-Choice Questions (cont.)

OBJECTIVES

- To understand that some multiple-choice questions involve reading a passage before answering
- To become familiar with reading numbered sentences in a passage
- To use knowledge of adjectives to answer questions

Teach/Model Tell students that some tests ask them to read a passage and then to answer questions about it. Read aloud the test tip, the directions, and the passage. Explain that the sentences are numbered because the numbers are used in the questions. Read aloud the first question. Model returning to the passage to find sentence 3. Then model selecting the correct answer and filling in the Answer Sheet.

Guided Practice Read aloud Question 2. Point out that the question number is 2, but the question asks about sentence 4 in the passage.

Independent Writing Practice Have students complete Questions 2 and 3 independently.

LESSON 5: WRITING TEST PRACTICE

Name ______________________

Writing Test Practice

B. Miranda's teacher asked the students in her class to write a paragraph about a family trip. Here is a first draft of Miranda's paragraph.

Read the test tip. Then answer questions 1–3 on the Answer Sheet below.

(1) I love to go camping with my family. (2) We go to the state park. (3) It is several miles upstate. (4) Last weekend we had the most fun. (5) At night we made a big campfire. (6) We told stories that were scary.

1. In sentence 3, which is an adjective that describes *how many*?
- A It
- B is
- C several
- D upstate

2. In sentence 4, which is an adjective that compares?
- F weekend
- G had
- H the
- I most

3. In sentence 5, which is an adjective that describes *what kind*?
- A at
- B night
- C made
- D big

Test Tip: There are many types of adjectives. Some adjectives tell *what kind* (small; red). Some adjectives tell *how many* (some; few). Some adjectives compare (older; oldest).

Answer all test questions on this Answer Sheet.

1. Ⓐ Ⓑ ● Ⓓ (C filled)
2. Ⓕ Ⓖ Ⓗ ● (I filled)
3. Ⓐ Ⓑ Ⓒ ● (D filled)

125 Writer's Companion ▪ UNIT 4
Lesson 5 *Writing Test Practice*

ASSESSING STUDENT RESPONSES

If a student answers test items incorrectly, consider possible reasons for the incorrect responses. One cause might be a failure to read the question carefully. Ask the student to walk you through the answering process. Then you will be able to determine why the student is missing important elements of the question. Is the student reading too quickly? Is the student unable to read some of the academic words in the question? Once you have determined the reasons for the incorrect responses, you will be able to guide the student to carefully read the questions and the answer choices.

Answering Multiple-Choice Questions (cont.)

OBJECTIVES

- To practice completing multiple-choice questions that involve reading a passage before answering
- To practice identifying errors in a passage and selecting the best answer

Teach/Model Tell students that some tests ask them to first read a passage and then to answer questions about it. Read aloud the test tip, the directions, and the passage. Explain that the sentences are numbered because the numbers are used in the questions that follow the passage. Read aloud the first question. Model returning to the passage to find sentence 1. Then model selecting the correct answer and filling in the Answer Sheet.

Guided Practice Read aloud Question 2. Point out that the question number is 2, but the question asks about sentence 3 in the passage.

Independent Writing Practice Have students answer Question 3 independently.

LESSON 5: WRITING TEST PRACTICE

Name ______________________

Writing Test Practice

C. Read the introduction, test tip, and the passage that follows. Then read each question and fill in the correct answer on your Answer Sheet.

James is in the third grade. He wrote this report about his brothers. He wants you to help him revise and edit the report. Read James's report and think about the changes he should make.

My Big Brothers

(1) I have two brothers. (2) I am the youngest one. (3) It is fun having brothers who are oldest. (4) They teach me lots of things. (5) Justin is teaching me to play drums. (6) He is an good drummer. (7) Richie helps me with my homework. (8) He is also teaching me chess.

1. What is the **BEST** way to revise sentence 1?

A I have three brother.

B I have two brother.

C I have one brothers.

D No revision is needed.

2. What change, if any, should be made in sentence 3?

F Change **brothers** to **brother**

G Change **brothers** to **a brothers**

H Change **oldest** to **older**

J Make no change

3. What change, if any, should be made in sentence 6?

A Change **good** to **gooder**

B Change **an** to **a**

C Change **good** to **best**

D Make no change

Test Tip:
Make sure that an adjective that tells *how many* matches its noun.
CORRECT: four books; one book
INCORRECT: four book; one books

Answer all test questions on this Answer Sheet.

1. Ⓐ Ⓑ Ⓒ Ⓓ (D filled in)
2. Ⓕ Ⓖ Ⓗ Ⓘ (H filled in)
3. Ⓐ Ⓑ Ⓒ Ⓓ (B filled in)

Writer's Companion ■ UNIT 4
Lesson 5 *Writing Test Practice*

126

Reaching All Learners

BELOW LEVEL

Have students restate in their own words what each test item is asking. For example, for Question 1, "I am looking for the best way to rewrite sentence 1." Then have partners talk through the steps they must follow to complete each question.

ADVANCED

Allow students to complete and check all items independently. Then have them compare their work with a partner.

ENGLISH LANGUAGE LEARNERS

Point out the words *if any* in Questions 2 and 3. Explain that the fourth choice in both of these questions is *Make no change*. Then work together to find the correct answers for Questions 2 and 3.

Answering Multiple-Choice Questions (cont.)

OBJECTIVES

- To become familiar with test items that require filling in a blank
- To understand that some writing tests contain multiple-choice questions

Teach/Model Tell students that some tests ask them to read a passage and then answer questions about it. Read aloud the test tip. Remind students to use their knowledge of comparisons to help answer the questions. Read aloud the directions and the passage. Explain that the blanks in the sentences are numbered because the numbers are used in the questions. Read aloud the first question. Model returning to the passage to find blank 1. Then model selecting the correct answer and filling in the bubble beside the question.

Guided Practice Read aloud Question 2. Remind students to carefully review each possible answer choice before selecting their final answer.

Independent Writing Practice Have students answer Questions 3–4 independently.

LESSON 5: WRITING TEST PRACTICE

Name ______________________________

Writing Test Practice

D. **Read "Spelling." Then read the test tip. Choose the word or words that correctly complete questions 1–4. Fill in the circle next to the correct answer.**

Spelling

I am good at spelling, but my sister Jenny is __(1)__. She has won many contests. She can spell words with many syllables. Next month she will enter __(2)__ big contest in the city. Then she will go on to the state championship. Maybe she will be __(3)__ famous than I will be. I help her learn __(4)__ additional word every day.

1. Which answer should go in blank (1)?
- Ⓐ speller
- Ⓑ gooder
- Ⓒ better

2. Which answer should go in blank (2)?
- Ⓕ a
- Ⓖ an
- Ⓗ those

3. Which answer should go in blank (3)?
- Ⓐ most
- Ⓑ better
- Ⓒ more

4. Which answer should go in blank (4)?
- Ⓕ a
- Ⓖ an
- Ⓗ more

Test Tip:
When comparing two things, use a different word than when comparing more than two things.
two: worse; redder
more than two: worst; reddest

127 **Writer's Companion** ■ UNIT 4
Lesson 5 *Writing Test Practice*

USING ACADEMIC LANGUAGE

Students will usually find academic language on tests. This is not language used in conversation but a specific type of language needed to understand directions, questions, and answer options.

Prepare students by helping them recognize and understand important words and phrases in directions. Explain that the word *choose* means to pick the best answer.

Look at Topic and Notes

OBJECTIVES

- To understand the terms *topic*, *purpose*, and *audience*
- To identify a topic in literature

Teach/Model After reading aloud the introduction, explain that a writer must consider several things before deciding what to write about. Choosing a *topic* involves thinking about a *purpose* for writing and about the *audience* (readers). Once chosen, the topic will guide you in doing research, note taking, writing, and revising. Read aloud the directions for Part A and the passage from *Leah's Pony*. As students listen, ask them to focus on any words from the title that are repeated in the passage. This will help them identify the topic of the passage.

Guided Practice After reading aloud the directions to Part B, guide students in completing Items 1–2. Ask students to name and underline the title of the book (Item 1). Then have them name words from the title that are in the passage as a way to identify the overall topic (Item 2).

Independent Writing Practice After reading aloud the directions, have students complete Part C independently.

Use a Literature Model

LESSON 1: TOPIC AND NOTES

Name ______________________

Writer's Craft in Literature

Look at Topic and Notes

What do you do first when you get ready to write? You choose a **topic,** which is what you will write about. You think about the **purpose,** which is the reason for writing. You also think about the **audience,** or who will read your writing. Your topic should not be longer than a few words. It guides you in doing research, taking notes, writing, and revising.

A. Read the following Literature Model. Think about the writer's topic.

Literature Model

The year the corn grew tall and straight, Leah's papa bought her a pony. The pony was strong and swift and sturdy, with just a snip of white at the end of his soft black nose. Papa taught Leah to place her new saddle right in the middle of his back and tighten the girth around his belly, just so.

—from *Leah's Pony*
by Elizabeth Friedrich

B. Read to figure out the topic.

1. Underline the title of the book.
2. Circle words from the title that you see in the paragraph.

C. What is the topic of the paragraph? How did the title and repeated words help you figure it out? Write your answer on the lines.

Possible response: The topic is Leah's pony. I saw the words *Leah* and *pony* in the title. Then I saw these words repeated in the paragraph. These clues helped me figure out the topic.

____

Writer's Companion • UNIT 5
Lesson 1 *Topic and Notes* 128

SHARING AND DISCUSSING

Have students compare the words they found in the passage that led them to its topic. Ask how the words lead to that conclusion, and what sort of words provide a hint that the topic is being discussed.

Explore Topic and Notes

OBJECTIVES

- To deepen students' understanding of how choosing a topic enhances writing
- To identify ways of choosing a topic efficiently
- To explore the process of narrowing a big idea into a chosen topic

Teach/Model Read aloud the introduction and review the diagram at the top of the page. Explain that this diagram shows one way of choosing a topic for a research report. Tell students that a carefully chosen topic makes writing easier and allows the writer to effectively reach the audience.

Guided Practice Read aloud the directions for Part A and model circling the word that tells which topic belongs on the line. Guide students in completing Item 1 by talking out the process of narrowing down the big idea of *water* to a specific topic. In order to find the correct topic, ask students *Which is most closely related to the ocean? A starfish or rain?* Repeat this process for Item 2 as well.

Independent Writing Practice After reading aloud the directions, have students complete Part B independently. For Part C, see Extending the Concept.

LESSON 1: TOPIC AND NOTES

Name ______________________

A Closer Look at Writer's Craft

Explore Topic and Notes

You can start thinking about your topic with a very big idea. Then you can narrow the idea a little bit. Then you can narrow it more. Then you'll have your topic!

A. Read the information in each triangle. Then circle the topic that belongs on the line.

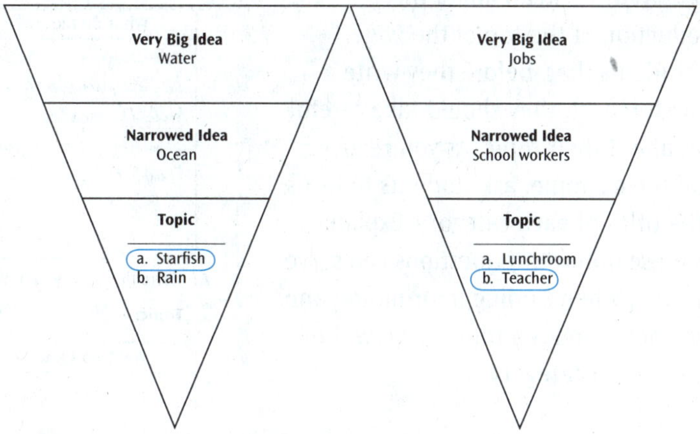

B. Read the very big idea and the narrowed idea. Then write a possible topic.

Very Big idea: Seasons Narrowed idea: Winter Topic: Possible response: snow

C. Prepare to write a research report. Write a very big idea that interests you. Then write a narrowed idea. Finally, write the topic for your research report.

Very big idea: Answers will vary, but should show a clear understanding of the

Narrowed idea: broadest topic as the **very big idea,** a narrower topic as the **narrowed**

Topic: **idea,** and the narrowest topic as the **topic.**

129 Writer's Companion • UNIT 5 Lesson 1 *Topic and Notes*

EXTENDING THE CONCEPT: TOPIC SENTENCES

Before students complete Part C, have them take another look at the Literature Model (page 128). Reread the passage aloud. Point out to students that there is no single sentence that declares what the topic is. Explain that this is often the case with fiction, in which the writer may want to write with more subtlety. Guide students in understanding that in nonfiction, such as their own research report, the writing often begins with a topic sentence. This sentence states clearly what the topic of the report will be. Encourage them to frame their own report topic as a topic sentence after completing Part C.

Use Topic and Notes

OBJECTIVES

- To prepare to write by preparing to take careful notes
- To practice taking notes about a chosen topic

Teach/Model Read aloud the introduction at the top of the page. Tell students that, before they write a research report, they should take careful notes about their topic. As you read aloud the example, ask students to focus on the titles of each category. Explain that these important questions can serve as a guide when finding information and taking notes about a topic. Review the notes in each category.

Guided Practice Read aloud the directions for Part A. Then have students write the topic of their research report on the topic line. Guide students as they fill out their charts with information that will prepare them to write about their chosen topics.

Independent Writing Practice After reading aloud the directions, have students complete Part B independently.

LESSON 1: TOPIC AND NOTES

Name ____________________

Practice with Writer's Craft

Use Topic and Notes

What do you do after you choose the topic? You prepare to take careful notes. This chart shows how one third grader used important questions to prepare.

Example Topic: *How Ants Work and Live*

What do I know?	What do I want to learn?	Where will I look?
1. Ants live in hills. *2. They build the hills themselves.* *3. They can carry heavy things.*	*1. How do ants work together?* *2. How do they get their food?* *3. Why don't they drown in their hills when it rains?*	*encyclopedias* *nature magazines* *science Web sites*

A. Write the topic you have chosen for your research report. Then complete the chart.

Topic: Answers will vary, but should be reasonable and well thought out.

What do I know?	What do I want to learn?	Where will I look?

B. Use another sheet of paper. Write each of the questions from the "What do I want to learn?" part of your chart. Tell where you will look for answers to each of the questions. Use information from the "Where will I look?" part of your chart to help you.

Writer's Companion • UNIT 5
Lesson 1 *Topic and Notes*
130

Reaching All Learners

BELOW LEVEL

Help students who are writing below grade level by making sure they have selected an appropriate topic. Encourage them to choose something they know. Then guide students to follow the model chart but complete it with their own details.

ADVANCED

Encourage students writing above grade level to take better notes. Challenge them to think about what they want their report to say and to consider in advance which details they will need in order to make this point.

ENGLISH LANGUAGE LEARNERS

Pair less fluent English speakers with more fluent English speakers. Have students discuss their topics and then help each other identify the appropriate English words and phrases to include in their charts.

The Parts of Notes

OBJECTIVES

- To understand how to take careful notes
- To analyze a Student Model

Teach/Model Read aloud the introduction and explain that a third grader took notes on the article "How Do Ants Live?" Point out that the student did a good job but not a perfect one. Read aloud the article and the steps for how to take notes. Then read aloud the model. As students listen, ask them to notice which details the student pulled out of the article.

Guided Practice Point out the items below the Student Model. Read aloud the first item. Then guide students in identifying and underlining the sentence that does not belong. Explain that this sentence does not belong because it is an opinion.

Independent Writing Practice Have students complete Items 2–3 independently. Then discuss their responses as a class.

Use a Student Model

LESSON 1: TOPIC AND NOTES

Name ________________

Focus on the Writing Form

The Parts of Notes

Careful **notes** tell where you found your information. They tell more about the topic. They include only facts, not opinions. The boxes below show an article and the notes a third grader took while reading the article. Read the article and then take notes. Then answer the questions.

How Do Ants Live?

We talked to a local scientist who studies ants. She said they are the most amazing creatures on the planet. She explained that ants use their jaws as they cut leaves. They take the leaves to their hills. They grow a special kind of food on the leaves.

Some kinds of ants make caves and nests deep under the ground. The ants can stay safe and keep their food there. They build in a special way so they won't drown while underground in the rain. You can read more about this on the next page.

People sometimes collect ants. They like to watch the ants march.

Student Model DRAFT

Topic: How Ants Work and Live
Notes from "How Do Ants Live?"
Ants are the most amazing creatures on Earth.
ants' jaws cut leaves
Ants get leaves to take home.
They grow a special kind of food on the leaves.
Some kinds of ants make caves and nests deep under the ground.
The ants can stay safe and keep their food there.

Name where the information was found.

Jot notes. Not all notes must be in complete sentences.

Include only **information that tells more about your topic.**

Only facts, not opinions, should be included.

Use your **own words** to take notes.

1. Underline a sentence that does NOT belong in the notes because it tells an opinion.
2. Circle information that is NOT written in the student's own words.
3. Draw two lines under notes the student has jotted in phrases, not complete sentences.

Writer's Grammar
Place quotation marks around the title of an article.

131 Writer's Companion ▪ UNIT 5
Lesson 1 *Topic and Notes*

WRITER'S GRAMMAR: QUOTATION MARKS

Help students notice that the title of the article needs to be set in quotation marks. Explain that quotation marks are used to set something apart. Tell students that, when they are writing about a magazine article, they should put the title in quotation marks. Point out the quotation marks missing in the article's title. Tell students to place quotation marks around the title of the article. Explain that there are different rules for punctuation for different kinds of printed material.

- *A Novel's Title*: italicized
- "A Poem": quotation marks

Evaluate Notes

OBJECTIVES

- To understand what to look for when evaluating notes
- To evaluate notes
- To suggest improvements for a set of notes

Teach/Model Using the points at the top of the page, discuss what to look for in evaluating a set of notes. Tell students that they will use these questions as a guide in evaluating the Student Model.

Guided Practice Read aloud the directions for Part A. Guide students when writing their responses and remind them to refer to the notes in the Student Model for information. Then point out the checklist in Part B and explain that it contains key things to keep in mind when evaluating notes.

Independent Writing Practice After reading aloud both sets of directions, have students complete Parts B and C independently.

LESSON 1: TOPIC AND NOTES

Name ______________________

Evaluating the Student Model

Evaluate Notes

How do you decide if someone has done a good job taking notes? Ask yourself these questions:

- Do the notes include only facts that tell more about the topic? (Read the topic. Then ask yourself what the notes explain about the topic. Think about whether the information states fact or opinion.)
- Does the writer use his or her own words? (Check the notes against where the information was found.)

A. Reread the Student Model on page 131. Then answer these questions.

1. Write two sentences or phrases that seem to be in the student's own words.

   ants' jaws cut leaves; Ants get leaves to take home.

2. Why did the student not include information about ant collectors in the notes?

   because the information does not tell more about the topic

B. Evaluate the Student Model. Check the box next to each thing the writer has done well. Do not check the box if the writer did not do a good job with something.

- ☐ The notes tell where the information was found.
- ☐ Only information that tells more about the topic is included.
- ☐ The notes include facts, but not opinions.
- ☐ The notes are written in the student's own words.
- ☐ Some of the notes are jotted, but are still clear.

C. How do you think the writer could take more careful notes? Write your ideas on another sheet of paper. Possible response: The writer should not copy exactly and should leave out opinions. The writer would have to think more about the information to keep from copying it.

Writer's Companion ■ UNIT 5
Lesson 1 *Topic and Notes*

132

See the rubric on page 207 for another way to evaluate the Student Model.

EXTENDING THE CONCEPT: TRANSITION WORDS

Point out to students that, in many cases, the notes they take will require them to describe a process with many steps. Explain that, when doing this, they should use transition words. Guide them in understanding that transition words like *first, next,* and *last* bring clarity to the steps in a process. Point out that these words usually appear at the beginning of a sentence or phrase. Have students work in pairs to brainstorm lists of possible transition words. Then have volunteers share their lists with the class.

Revise by Paraphrasing Information

OBJECTIVES

- To revise notes by paraphrasing information
- To implement improvements in a set of notes

Teach/Model Tell students that, in this activity, they will revise a set of notes by paraphrasing.

Guided Practice Read aloud the introduction and the example. Discuss with students how paraphrasing helps a note-taker find the most important information in their research material. This information is made understandable by writing it in their own words. Read aloud the directions for Part A. Then model how you would revise Item 1. Guide students in writing phrases in their own words while completing Item 2.

Independent Writing Practice After reading aloud the directions, have students complete Part B independently. Invite volunteers to share their revisions aloud.

LESSON 1: TOPIC AND NOTES

Name ______________________

Revising the Student Model

Revise by Paraphrasing Information

What is one thing the student writer could have done better? The student could have paraphrased while taking notes. **Paraphrasing** means "writing in your own words." Here is an example of how a sentence from the Student Model could be improved.

Example They grow a special kind of food on the leaves.

They use leaves as a place to grow food.

A. Find two other sentences in the Student Model that are NOT written in the student's own words. First, write each sentence as the student wrote it. Then write the sentence in your own words. Answers will vary. Possible responses are given.

1. Sentence from Student Model: Some kinds of ants make caves and nest deep under the ground.

   A better way to write the notes: Answers should vary, but should include same information in student's own words.

2. Sentence from Student Model: The ants can stay safe and keep their food there.

   A better way to write the notes: Answers will vary, but should include same information in student's own words.

B. On another sheet of paper, write the research topic you chose earlier. Begin your research by taking careful notes. Use as many sheets of paper as you need.
Notes will vary, but should be thorough. Notes should be in student's own words and should not include opinions.

WRITER'S STRATEGY: NOTE TAKING

Explain to students that taking notes is an important part of the writing process. Tell them that their notes will be useless if they are not written clearly. On the board, outline some things to keep in mind when taking notes:

- Write neatly.
- Keep notes from the same source together.
- Paraphrase but include enough information so that it is understandable later.

Look at Organizing Information

OBJECTIVES

- To understand the terms *main idea and details, time order,* and *compare and contrast*
- To identify organizing information as a step in the writing process

Teach/Model Explain that, after choosing a topic and taking notes, a writer must put his or her ideas in careful order. Tell them this is called *organizing information.* Go on to explain that there are many ways of organizing information. Explain that a writer should choose the organizing method that suits his or her topic. Read aloud the introduction, the directions for Part A, and the passage from *Yippee-Yay!* Have students listen for any words that might tell the paragraph's main idea and details that describe that main idea.

Guided Practice Ask students to name some of the words that helped them find the main idea and details of the paragraph. Then read aloud the directions for Part B and guide students in completing it.

Independent Writing Practice After reading aloud the directions, have students complete Part C independently.

Use a Literature Model

LESSON 2: ORGANIZING INFORMATION

Name ______________________________

Writer's Craft in Literature

Look at Organizing Information

What is the next step after you choose a topic and take notes? You organize your information. You write in a way that is interesting and makes sense. There are many ways to organize information. Here are three of the ways.

- Main idea and details: most important ideas with small pieces of important information to explain more about them
- Time order: events in the order they happened
- Compare and contrast: how things are alike and different

A. Read the following paragraph. Notice how the writer has organized information.

Literature Model

Roping was the most difficult skill for a cowboy to learn, and it was the most important. Cowboys carried ropes called lariats to lasso cattle. It took many months of practice to learn to spin the lariat and release it at just the right moment.

—from *Yippee-Yay!*
by Gail Gibbons

B. Find the words that tell the main idea and details.

1. Circle the main idea.
2. Underline the sentences that tell important information about the main idea.

C. What kind of organization did the author use?

main idea and details

Writer's Companion ■ UNIT 5
Lesson 2 *Organizing Information* 134

SHARING AND DISCUSSING

Pair students and have them compare their notions of what the main idea is in the Student Model. Ask how they reached their conclusions and how the organization of the paragraph helped them find the main idea. After students have shared with their partners, discuss as a class.

Explore Organizing Information

OBJECTIVES

- To deepen students' understanding of how organizing information enhances writing
- To identify ways to organize information
- To explore choosing the best way to organize information for a given topic

Teach/Model Use the introduction and diagram at the top of the page to help students understand three ways of organizing information.

Guided Practice Explain that writers organize their information in order to make their writing as clear as possible. Organizing writing in a way that suits the topic makes writing easier to understand. Read aloud the directions for Part A and model circling the words that tell how the paragraph is organized. Guide students in completing Items 1–2.

Independent Writing Practice After reading aloud the directions, have students complete Part B independently. For Part C, see Extending the Concept.

LESSON 2: ORGANIZING INFORMATION

Name ______________________________

A Closer Look at Writer's Craft

Explore Organizing Information

Think about the information you will want to include in your research report. Then think about which kind of organization will work best for your report.

Ways to Organize Your Research Report

- **Main Idea and Details:** important ideas with smaller pieces of information to tell more about them
- **Time Order:** events in the order they happened
- **Compare and Contrast:** how things are alike and different

A. Read each paragraph. Circle the answer that tells how it is organized.

1. Our library is filled with interesting books. They are grouped by type of book or subject. The library receives new books every month.

   a. main idea and details **b.** time order **c.** compare and contrast

2. Fall and winter are alike in some ways. They are different in others. There are cold days during both seasons. Leaves change colors in fall. The trees are bare in winter, though.

   a. main idea and details **b.** time order **c.** compare and contrast

B. Read this paragraph from *Yippee-Yay!* Circle the answer that tells how it is organized.

Next, the cowboys lassoed the calves. This was called chopping out. One by one, the calves were brought to a wood fire filled with heated branding irons. The cowboys took turns pressing a branding iron to each calf's hip, leaving a mark.

a. main idea and details **b.** time order **c.** compare and contrast

C. Choose one of the kinds of organization. Circle it below. Then use it to write a short paragraph on another sheet of paper. Answers will vary.

a. main idea and details **b.** time order **c.** compare and contrast

135 **Writer's Companion** ■ UNIT 5
Lesson 2 *Organizing Information*

EXTENDING THE CONCEPT: WORD CHOICE

Before students complete Part C, reread the Literature Model (page 134) aloud. Point out the writer's use of the word *most*. Explain that a word like *most*—especially when it modifies the word *important*—can be a signal that a piece of writing is organized by order of importance. Explain that order of importance is another type of organization. It lists details from least to most important or from most to least important. Guide students to see that this word is a superlative, and that there are many words of this kind that can come in handy when organizing by order of importance. Have students pair up to generate lists of superlatives and of words that compare. Model some of these words: *faster, fastest; zesty, zestiest.* Have a volunteer from each pair read their list for the class.

Use Organizing Information

OBJECTIVES
- To prepare to write by organizing information
- To practice organizing information about a topic

Teach/Model Tell students that, before they write a research report, they should spend some time organizing the information for their report. Read aloud the introduction and the example at the top of the page. In the example, have students notice that the choice of organization (main idea and details) led to the outline form in which the information is presented.

Guided Practice Read aloud the directions for Part A. Have students write the name of their topics for their research reports on the topic line. Then explain that they need to decide how to organize their information. Guide them in filling out their outlines with notes that will support their chosen topics.

Independent Writing Practice After reading aloud the directions, have students complete Part B independently.

LESSON 2: ORGANIZING INFORMATION

Name ______________________

Practice with Writer's Craft

Use Organizing Information

Prepare to write an outline that organizes information for a research report. First, decide how you will organize your report. This student decided to organize her outline by main idea and details.

Example **My topic:** *The Anasazi community*

How I will organize: *main idea and details*

I. Introduction
  A. Anasazi were Native Americans.
  B. They settled long ago.

A. Think about a topic for a research report on Native Americans. Then write your topic. Tell how you will organize your report.

My topic: Answers will vary.

How I will organize: ______________________

B. Research your topic. Take careful notes. Then write the first part of an outline for your research report.

Title:

I. Introduction
  A. ______________________
  ____
  B. ______________________
  ____

Writer's Companion ▪ UNIT 5
Lesson 2 *Organizing Information* 136

Reaching All Learners

BELOW LEVEL

Help students who are writing below grade level to understand the parts of an outline. Ask where the most important information goes. Suggest students reword the Student Model with their main idea and details.

ADVANCED

Encourage students writing above grade level to include more details about their topic. Remind them to be careful when doing this and not to stray off topic or put things into their outline that will later need to come out.

ENGLISH LANGUAGE LEARNERS

Pair less fluent English speakers with more fluent English speakers. Have students discuss their topics and then help each other identify the appropriate English words and phrases to include in their outlines.

The Parts of an Outline

OBJECTIVES
- To understand how an outline is organized
- To analyze a Student Model

Teach/Model Tell students that the outline on this page was done by a third grader. Read aloud the introduction and the steps for how to organize an outline. Then read aloud the model. As students listen, have them follow along, paying attention to the topic and the outline format.

Guided Practice Point out the items below the Student Model. Read the first item aloud. Then guide students in identifying and drawing an arrow pointing to the topic of the outline.

Independent Writing Practice Have students complete Items 2–4 independently. Then discuss their responses as a class.

Use a Student Model

LESSON 2: ORGANIZING INFORMATION

Name ____________________

Focus on the Writing Form

The Parts of an Outline

An **outline** helps you organize ideas and information. Below is a draft of an outline written by a third grader. Look at the form of the outline. Note the information the writer included. Then answer the questions.

Student Model

DRAFT

The Anasazi Community ←
by Keisha

I. Introduction
   A. Anasazi were Native Americans.
   B. They settled long ago.
II. All the Anasazi helped with food.
   A. Men made things.
   B. Everyone farmed fields.
III. Home life was important.
   A. Men and women built homes.
   B. Women and children did things.
   C. Women and children made useful things.
IV. Conclusion
   A. The Anasazi became a strong community.
   B. They lived successfully for hundreds of years.

The introduction tells the main idea of the report.

Each main part of the outline includes a **Roman numeral followed by a period:** I. II. III. IV.

The outline includes a **capital letter followed by a period** to show each important detail: A. B. C. D.

In this outline, **the body (middle)** has facts organized under two main headings, II. and III.

Details are organized in an order that makes sense.

The conclusion restates and tells a bit more about the main idea from the introduction.

1. Draw an arrow to the topic of the outline.
2. Draw a box around the information in the introduction of the outline.
3. Circle the body of the outline.
4. Underline the conclusion in the outline.

Writer's Grammar
An outline has main topics and important details. Count how many topics are in this outline. Count how many important details there are.

137 Writer's Companion • UNIT 5
Lesson 2 *Organizing Information*

WRITER'S GRAMMAR: MAIN IDEA AND DETAILS

Draw students' attention to the Student Model, guiding them to see that it is made up of main topics that are supported by important details. Tell them that, in the model, there are four main topics, each supported by two or three important details. Point out that the number of details will vary depending on the complexity of a given main topic. Also, it is important, for clarity's sake, to keep the number of details as few as possible when making an outline.

Evaluate an Outline

OBJECTIVES

- To understand what to look for when evaluating an outline
- To evaluate an outline
- To suggest improvements for an outline

Teach/Model Using the points at the top of the page, discuss what to look for when evaluating an outline.

Guided Practice Read aloud the directions for Part A. Guide students in completing Items 1–2 and remind them that they must use the Student Model to find the answers. Then point out the checklist in Part B and explain that it will help them focus on the key parts of an outline. Tell students to keep it handy for when they write their outlines in the future.

Independent Writing Practice After reading aloud both sets of directions, have students complete Parts B and C independently.

LESSON 2: ORGANIZING INFORMATION

Name ______________________________

Evaluating the Student Model

Evaluate an Outline

Ask yourself these questions to figure out whether an outline is strong:

- Does the writer include an introduction, a body, and a conclusion? (Look for the words *Introduction* and *Conclusion*. The body is the main part.)
- Does the writer include main parts and important details? (Look for Roman numerals and capital letters.)

A. Reread the Student Model on page 137. Then answer these questions.

1. Think about the first paragraph that will come after the introduction. What will the paragraph be about?

   how all the Anasazi helped with food

2. What kind of organization did the writer choose? How can you tell?

   Possible response: The writer chose main idea and details. I can tell because the writer tells main ideas and then tells more about them.

B. Evaluate the Student Model. Check the box next to each thing the writer has done well. Do not check the box if you do not think the writer has done a good job.

- ☐ The writer included an introduction, body, and conclusion.
- ☐ The writer used Roman numerals to show main sections.
- ☐ The writer used capital letters before the important points.
- ☐ The details are organized in an order that makes sense.
- ☐ The conclusion restates and tells a bit more about the main idea of the report.

C. How do you think the writer could make the outline better? Write your ideas below. Use another sheet of paper if you need more space.

Possible response: The writer should tell more details. The writer should tell details that are clearer.

See the rubric on page 207 for another way to evaluate the Student Model.

EXTENDING THE CONCEPT: USING GRAPHIC ORGANIZERS

Suggest to students that for some people the Roman numerals and letters of an outline may be confusing or even a little intimidating. Explain that another way to rough out your ideas is to jot them into different sized boxes. Guide them to see that a larger box should be used to indicate a main topic, while the important ideas should go in smaller boxes below it. Have students work in pairs or small groups and use boxes to contain their suggested improvements for the Student Model.

Revise by Adding Details

OBJECTIVES

- To revise an outline by adding details
- To see how adding details can improve an outline
- To implement improvements in an outline

Teach/Model Tell students that they will be revising part of the Student Model by adding details. Explain that the purpose is to improve the outline.

Guided Practice Read aloud the introduction and the example. Discuss with students how adding specific facts and details enriches the outline. Read aloud the directions for Part A. On the board, model how you would revise the outline in Part A. Have students take turns describing what they would do to revise the outline.

Independent Writing Practice After reading aloud the directions, have students complete Part B independently. Invite volunteers to share their revisions aloud.

LESSON 2: ORGANIZING INFORMATION

Name ______________________

Revising the Student Model

Revise by Adding Details

The student could have added more details to make the outline better. Here is an example of how part of the Student Model could be improved by adding more details.

Example **Original:**

I. Introduction
   A. Anasazi were Native Americans.
   B. They settled long ago.

Revision:

I. Introduction
   A. Anasazi were Native Americans.
   B. They settled in the Southwest more than 1,000 years ago.
   C. Everyone in the community had a special job.

A. Tell what you would do to revise the following part of the outline. You do not need to rewrite the outline. Just tell the kinds of things you would do to change it.

II. All the Anasazi helped with food.
   A. Men made things.
   B. Everyone farmed fields.

Possible response: I would tell what things the men made and what crops were farmed. I might add a C. and maybe a D.

B. Revise the beginning of the outline you wrote on page 136. Then complete the outline. Write your outline on another sheet of paper.

Answers will vary, but should include correct outline format, strong organization, and plenty of details.

139 Writer's Companion ▪ UNIT 5
Lesson 2 *Organizing Information*

WRITER'S STRATEGY: TOPIC SENTENCE

Explain to students that a great topic sentence helps the reader focus. They also do something more—they pique the reader's interest in what is to follow. Read aloud this example: "Most of what we once knew about whales is wrong." Explain that the audience not only knows that they will hear about whales, but also that they will probably hear something new and interesting. Suggest that students try this with their own topic sentences. Then have them practice using the sentences in their drafts (next lesson).

Look at Writing a Research Draft

OBJECTIVES

- To understand the terms *introduction, middle,* and *end*
- To identify how a first draft is used in writing a research report

Teach/Model Tell students that, after choosing a topic, taking notes, and creating an outline, the writer of a report makes a first draft; the first draft is revised later. Go on to explain that a good first draft includes an introduction, a middle, and an end. Explain that the first draft is a rougher version of a finished research report. Read aloud the introduction, the directions for Part A, and the passage from *Boom Town.* Have students listen for any words that might tell the important idea of the passage.

Guided Practice Read aloud the directions for Part B. Ask students to share how they found the important idea of the passage. Then guide students in completing Part B.

Independent Writing Practice After reading aloud the directions, have students complete Part C independently.

Use a Literature Model

LESSON 3: WRITING A RESEARCH DRAFT

Name ______________________________

Writer's Craft in Literature

Look at Writing a Research Draft

You have chosen a topic, taken careful notes, and made an outline. What is your next step? You will use your outline to write a draft of your research report. Your outline will guide you as you think about the introduction, middle, and end.

A. Read the following paragraph. Think about an outline the writer might have used to write the paragraph.

Literature Model

It took us twenty-one days on the stagecoach to get to California. When we got there, I thought we'd live with Pa in the gold fields. A whole tent city was built up. But Ma shook her head. "The gold fields are no place for children. We'll get a cabin and live in town."

—from *Boom Town*
by Sonia Levitin

B. Identify details based on an outline.

1. Think about an outline the author might have created before writing the book. Circle the sentence that introduces the important idea.
2. Underline details that tell more about the important idea.

C. Write an outline you think the author might have used for this part of the book. Use another sheet of paper. Answers will differ, but should show correct form for the outline. Outlines should include main idea and details from the paragraph.

Writer's Companion • UNIT 5
Lesson 3 *Writing a Research Draft* 140

SHARING AND DISCUSSING

Have students compare the ways they found the important idea of the Literature Model. Ask if they needed to read the whole passage first and how they separated the important idea from the details.

Explore Writing a Research Draft

OBJECTIVES

- To deepen students' understanding of how solid outlines enhance writing
- To identify ways to use an outline to write a research report
- To explore choosing the best way to use an outline to write a research report draft

Teach/Model Read aloud the introduction. Using the diagram at the top of the page, explain to students how to convert an outline into a draft with a clear beginning, middle, and end.

Guided Practice Explain that writers create outlines to make their first draft as clear as possible. Tell students that using an outline to build a draft will also make the words come easier. Read aloud the directions for Part A and Item 1. Model underlining part of the sample outline that will become the beginning of the research report. Continue guiding students in completing Items 2–3.

Independent Writing Practice After reading aloud the directions, have students complete Part B independently.

LESSON 3: WRITING A RESEARCH DRAFT

Name ____________________

A Closer Look at Writer's Craft

Explore Writing a Research Draft

Follow the order of ideas in your outline as you write your **research report draft.** The main parts of your outline, I., II., III., IV., will become the main parts and main ideas of your report. The details from your outline, A., B., C., D., will become the details in each main part.

I. Introduction	becomes ➡	**Beginning** Introduction of your report
II. and III. Body	becomes ➡	**Middle** Body of your report
IV. Conclusion	becomes ➡	**End** Conclusion of your report

A. Read the parts of the outline. Tell what each part will become in the research report.

The Anasazi Community

I. Introduction
   A. Anasazi were Native Americans.
   B. They settled in the Southwest more than 1,000 years ago.
   C. Everyone in the community had a job.
II. All the Anasazi helped with food.
   A. Men hunted and made spears, bow, and arrows.
   B. Everyone farmed fields.
III. Home life was important.
   A. Men and women built homes.
   B. Women and children cared for households.
   C. Women and children made useful things.
IV. Conclusion
   A. The Anasazi became a strong community.
   B. They lived successfully for hundreds of years.

1. Part of the outline will become the beginning. Underline it.
2. Put a box around the part that will become the body.
3. Circle the part that will end the report.

B. Suppose you are ready to write about how the Anasazi will help with food. Which details from the outline will you put in your draft? Write them on another sheet of paper.

Possible response: Men hunted and made spears, bow, and arrows. Everyone farmed fields.

141 Writer's Companion ■ UNIT 5
Lesson 3 *Writing a Research Draft*

EXTENDING THE CONCEPT: ORDERING DETAILS

Before students complete Part B, reread these sentences from the Literature Model (page 140) aloud: *A whole tent city was built up. The gold fields are no place for children.* Explain that these are details that support the important idea that the narrator has traveled to a mining camp in California. Guide students to see that while these are supporting details, in this case there is a particular order in which the details should appear. Make it clear that order is not always an issue, but when things happen chronologically—or in time order—which detail comes first matters. Suggest students pair up to discuss whether or not any details in the Anasazi outline requires a particular order.

Use Writing a Research Draft

OBJECTIVES

- To prepare to write by using an outline
- To practice using an outline to write the first draft of a research report

Teach/Model Tell students that, before they write a research report, they should spend some time converting the important ideas and details of an outline into a topic sentence and details. Read aloud the introduction at the top of the page. Have students note that the words in the draft are not identical to the ones in the outline. Point out that the writer must combine some things and move others around to make the topic and details clear.

Guided Practice Read aloud the directions for Part A. Then have students copy the outline for their research report into the chart. Guide them in converting the outline into a topic sentence and details. Encourage students to use the topic sentences they created in Lesson 2.

Independent Writing Practice Have students complete Part B independently.

LESSON 3: WRITING A RESEARCH DRAFT

Name ____________________

Practice with Writer's Craft

Use Writing a Research Draft

Your outline will give you information to begin your research report. It will show you the main parts. It will also show you the details to include in each part. Here is how one third grader used an outline to begin a draft.

Example

Outline	Research Report Draft
I. Introduction A. Anasazi were Native Americans. B. They settled in the Southwest more than 1,000 years ago. C. Everyone in the community had a job.	**Topic Sentence:** The Anasazi were Native Americans who settled in the Southwest. **Details:** They settled there more than 1,000 years ago. All the people in the Anasazi community had their own special jobs to do.

A. Look back to the outline you started in Lesson 2. Copy it below. Then complete the chart. Use the important points to write a topic sentence. Then add details.

Outline	Research Report Draft
	Topic Sentence: **Details:**

B. Use information from your chart to write a paragraph that supports your topic sentence. Use another sheet of paper.

Writer's Companion • UNIT 5
Lesson 3 *Writing a Research Draft* 142

Reaching All Learners

BELOW LEVEL

Help students who are writing below grade level convert the words from their outlines into a topic sentence. Then have them compare their new topic sentences to the ones created in Lesson 2.

ADVANCED

Encourage students writing above grade level to include more details about their topic, or, if necessary, list those details in time order as they fill out the chart. Have them concentrate on basing their details on interesting facts.

ENGLISH LANGUAGE LEARNERS

Pair less fluent English speakers with more fluent English speakers. Have students discuss their topics and then help each other identify the appropriate English words and phrases to include in their charts.

The Parts of a Draft

OBJECTIVES
- To understand how a research report is organized
- To analyze a Student Model

Teach/Model Explain to students that the writing model on this page is a first draft of a third grader's research report. Read aloud the introduction and the steps for how to organize a research report. Then read aloud the model. As students listen, have them pay careful attention to how the main topics are supported by details.

Guided Practice Point out the items below the Student Model. Read aloud the first item. Then guide students in identifying and drawing a box around the topic sentence of the whole report.

Independent Writing Practice Have students complete Items 2–4 independently. Then discuss their responses as a class.

Use a Student Model

LESSON 3: WRITING A RESEARCH DRAFT

Name ______________________

Focus on the Writing Form

The Parts of a Draft

The main purpose of your **draft** is to present facts in a way the reader will understand. Your draft is not supposed to be a perfect piece of writing. There may be some errors in grammar or sentence order. Some important sentences may be left out. Here is part of a draft written by a third grader. Read to figure out if the writer has achieved the main purpose. Then answer the questions.

Student Model

DRAFT

The Anasazi Community
by Keisha

The Anasazi were Native Americans who settled in the Southwest. All the people in the Anasazi community had their own special jobs to do.

Home life to the Anasazi was very, very important. Men and women built their homes in the Cliffs on top of mesas. I wouldn't want to live up so high. The Anasazi settled in the Southwest more than 1,000 years ago. women and childern took care of the households. Women and children also made clothing, pots, and many other items.

The report has a **title.**

The topic sentence of the report tells the main idea of the whole report.

Each paragraph includes a topic sentence to tell the main idea of the paragraph.

The writer follows the **outline.**

The writer includes **details and examples** from the outline.

The report includes only **facts** about the topic.

1. Draw a box around the topic sentence of the whole report.
2. Circle the topic sentence of the second paragraph.
3. Underline the details in the second paragraph that support its topic sentence.
4. Compare the outline and the draft. Which sentence is out of order?

The Anasazi settled in the Southwest more than 1,000 years ago.

Writer's Grammar
Check to see that you have started each sentence with a capital letter. Find a sentence in the draft where the writer forgot to do this.

143 Writer's Companion ▪ UNIT 5
Lesson 3 *Writing a Research Draft*

WRITER'S GRAMMAR: CAPITALIZATION

Draw students' attention back to the Student Model and point out where the writer failed to capitalize the first word of the next to last sentence. Explain that the first word of every sentence must be capitalized. Ask students to volunteer examples of other situations in which words must be capitalized. Guide them to see that proper nouns, like the names of people, are one such example. Go on and explain that titles of books and other publications must also begin with a capital letter.

Evaluate a Draft

OBJECTIVES

- To understand what to look for when evaluating a draft
- To evaluate a draft
- To suggest improvements for a draft

Teach/Model Discuss with students why it is important to evaluate writing. Explain that writing is rarely finished after the first draft, so students should assume they will evaluate and revise drafts in order to improve them. After reading aloud the points at the top of the page, discuss what to look for when evaluating a draft.

Guided Practice Read aloud the directions for Part A. Guide students in writing their responses and referring to both the Student Model outline and draft for information. Tell students that using a checklist is helpful in evaluating writing because it allows one to focus on a few important elements. Point out the checklist in Part B and discuss the key items to evaluating a draft.

Independent Writing Practice After reading aloud both sets of directions, have students complete Parts B and C independently. Have volunteers share their suggestions with the class.

LESSON 3: WRITING A RESEARCH DRAFT

Name ______________________

Evaluating the Student Model

Evaluate a Draft

Ask yourself these questions to decide whether the draft is strong:

- Does the writer follow the points in the order in which they appear in the outline? (Compare the outline to the draft. Look for the specific main ideas and details.)
- Does each paragraph include a topic sentence with details to tell more about it? (Look for the topic sentence. Note the information in each detail.)

A. Reread the Student Model on page 143. Then answer these questions.

1. Which detail sentence tells where the Anasazi built their homes?

   Men and women built their homes in the Cliffs on top of mesas.

2. Which sentence does not belong in the second paragraph because it does not tell a fact?

   I wouldn't want to live up so high.

B. Evaluate the draft by referring back to page 143. Check the box next to each task the writer has done well. Do not check the box if the writer has not done a good job.

- ☐ The writer used ideas from the outline.
- ☐ The writer followed the order of ideas from the outline.
- ☐ The writer included only facts about the topic.
- ☐ The writer wrote a topic sentence for each paragraph.
- ☐ The writer followed the outline to add facts and examples to support each topic sentence.

C. How do you think the writer could make the draft better? Write your ideas below.

Possible response: The writer should write sentences in order. The writer should include only facts. The writer should include more details.

See the rubric on page 207 for another way to evaluate the Student Model.

EXTENDING THE CONCEPT: WORD CHOICE

Explain to students that, as they turn the main topics and details of an outline into a first draft, word choice becomes quite important. Read aloud this sentence from the Student Model: *Home life to the Anasazi was very, very important.* Point out that the words *very, very,* while clear in their meaning, are not very expressive. Guide them in using a thesaurus in order to find better words like *highly* and *hugely*.

Revise by Adding a Bibliography

OBJECTIVES

- To revise a research report draft by adding a bibliography
- To see how a bibliography improves a research report
- To implement improvements in bibliography entries

Teach/Model Tell students that, in this activity, they will revise a draft by adding a bibliography.

Guided Practice Read aloud the introduction and the example. Discuss with students how the bibliography entry was changed and improved. Read aloud the directions for Part A and Item 1. On the board, model revising Item 1 as a class. Then have students write out their own revisions for Item 2.

Independent Writing Practice After reading aloud the directions, have students complete Part B independently.

LESSON 3: WRITING A RESEARCH DRAFT

Name ________________________________

Revising the Student Model

Revise by Adding a Bibliography

The writer could have done a better job with the draft by including a bibliography. A bibliography lists the sources used for research. Each source is called an entry.

- The entries are written in alphabetical order based on the author's last name.
- Book titles are underlined or typed in *italics*.

Here is an example of how a bibliography entry can be improved. Note where the commas and periods have been placed in the revised entry.

Example **Book Entry**

Before Revision:	After Revision:
lopez, Rhonda, History of the anasazi. New York: Fine Books, 2004	Lopez, Rhonda. History of the Anasazi. New York: Fine Books, 2004.

Labels: Author; Book title; Location and name of publisher; Date the book was published

A. **Read the bibliography entries below. Revise each one. Make sure capitalization, punctuation, and form are correct.**

1. Neely, Pat. life in the Southwest. New York: Excellent Books, 2004.

   Revised entry: Neely, Pat. *Life in the Southwest.* New York: Excellent Books, 2004.

2. ali, John, Native americans of California. Chicago, Reigel, 2003.

   Revised entry: Ali, John. *Native Americans of California.* Chicago: Reigel Books, 2003.

B. **Return to the draft you began on page 142. On another sheet of paper, write two bibliography entries for resources you used.**

Answers will vary, but should include correct form and punctuation.

145 Writer's Companion • UNIT 5 Lesson 3 *Writing a Research Draft*

WRITER'S STRATEGY: CHOOSING SOURCES

Explain that a research report is only as good as the sources that are used to write it. Tell students that standard sources include encyclopedias, history and science books, actual documents (primary sources), and journal articles. Make it clear that while some of these sources are available on the Internet, students must be careful in avoiding articles that are unsourced, or written by nonexperts. Suggest that they use the Internet as a guide to printed sources.

Review Organization

OBJECTIVES

- To review the terms *topic, notes, outline,* and *draft*
- To identify how an outline is turned into a draft

Teach/Model Review with students that writers build research reports and other writing by organizing their work through a series of steps. Explain that a writer first chooses a *topic* and takes *notes*. He or she then creates an outline and turns that *outline* into a *draft*. Explain that following these steps will lead to clearer, more effective writing. Read aloud the introduction, the directions for Part A, and the passage from *Cocoa Ice*. As students listen, ask them to pay attention to the words that hint at the type of organization that is being used in the passage.

Guided Practice After reading aloud the directions, ask students to identify the phrases in the passage that helped them see how the writer organized it. Point out that clear organization is key for getting a message across. Then guide students in completing Part B.

Independent Writing Practice After reading aloud the directions, have students complete Part C independently.

Use a Literature Model

LESSON 4: REVIEW ORGANIZATION

Name ______________________

Writer's Craft in Literature

Review Organization

When you prepare to write a research report, you start by choosing a **topic.** Next, you take careful **notes.** You then use your notes to make an **outline.** The information in your outline will guide you in writing a **draft** of your report.

A. Read the paragraph. Notice how the writer has organized information.

Literature Model

Little pink cacao flowers grow right on the trunk. Green cacao pods grow side by side with the flowers, and next to them grow ripe yellow and red pods, ready to be picked. A cacao tree is always blooming, always ripening, and always ready to harvest.

—from *Cocoa Ice*
by Diana Appelbaum

B. Read the paragraph again. Think about how it is organized.

Possible responses given.

1. Underline the phrases that best help you figure out the topic of the paragraph.
2. Circle the kind of organization the writer uses in the paragraph.

a. main idea and details **b.** time order **c.** compare and contrast

C. Write a phrase that tells the topic of the paragraph.

Answers will vary, but should be a phrase from the Literature Model that relates to the topic of a cacao tree.

SHARING AND DISCUSSING

Have students compare the topic phrases they found. Guide them to see that, when a piece of writing is organized by main idea and details, the topic sentence will not always come first. Invite students to consider that if the last sentence of the passage came first, would it make a clear topic sentence?

Review Organization

OBJECTIVES

- To review the steps of organizing writing
- To review narrowing a topic

Teach/Model Read aloud and review the introduction and graphic organizer at the top of the page. Explain the different steps in organizing writing: choosing a topic, taking notes, making an outline, and writing a draft.

Guided Practice Review that writers of research reports begin with a big idea and then narrow it down to a manageable size. Remind students that writers write with a point in mind. Read aloud the directions for Part A and model how to narrow a topic. Guide students in completing Items 1–2. Review responses as a class.

Independent Writing Practice After reading aloud the directions, have students complete Part B independently.

LESSON 4: REVIEW ORGANIZATION

Name ______________________

A Closer Look at Writer's Craft

Review Organization

In this unit, you learned these steps to use when you organize.

1. Choose a **topic.** Be sure to narrow it for your writing.
2. Take **careful notes**. Write to tell where you find information. Write facts that tell more about your topic.
3. Use your notes to make an **outline.** Include main ideas and supporting details in an order that makes sense.
4. Use your outline to write a **draft** of your research report. Include the same main ideas and details. Your draft does not have to be perfect.

A. **Look at the graphic organizer. Write the answers to the questions.** Possible responses given.

1. Why should you narrow your topic?

   A broad topic will be impossible to research corrrectly.

2. On what information are you taking notes?

   on information I looked up in books or on the web

B. **Read these big ideas. Narrow each one to make a strong research report topic.**

Example **Idea:** planets **Topic:** *rings of Saturn*

1. **Idea:** weather **Topic:** Possible response: hail
2. **Idea:** cars **Topic:** Possible response: hybrid cars
3. **Idea:** birds **Topic:** Possible response: the Dodo bird

147

Writer's Companion ▪ UNIT 5
Lesson 4 *Review Organization*

EXTENDING THE CONCEPT: CHOOSING A TOPIC

Before students complete Part B, suggest that they make sure that the topic they choose is something they are really interested in. Guide them to see that the more they know—or care about—the subject of a research report, the easier it will be to write. Point out that the writing will also be clearer and more exciting for the reader. Suggest they choose one of the "big ideas" that appeals to them and to then narrow it down to an even more interesting idea.

Review Organization

OBJECTIVES

- To practice a prewriting technique
- To review organizing notes
- To determine the type of structure suggested by a set of notes

Teach/Model Tell students that, before they write a research report, they should decide on a structure for their writing. They should also figure out how they will turn their notes into an outline. Read aloud the text at the top of the page and review with students the example. Discuss the notes and how they are organized.

Guided Practice Read aloud the directions for Part A. Then guide students in determining the type of "big idea" the writer used. Continue guiding students as they complete Part A.

Independent Writing Practice After reading aloud the directions, have students complete Part B independently.

LESSON 4: REVIEW ORGANIZATION

Name ____________________

Practice with Writer's Craft

Review Organization

Organization is important as you plan your report. The example in the box shows some of the notes a third grader wrote while planning a report.

Notes

Bees get their food from flowers.
Bees keep their food in the hive. They can eat it when it's too cold to fly to find more.
Young bees get their food from worker bees.
After bees find food, they do a kind of dance to show other bees where to go back to look for more.

A. Read the student's notes. Answer questions about the organization. Possible responses given.

1. What big idea do you think the student writer had before choosing a topic? Why do you think so?

   I think the writer's big idea was probably bees. All the notes tell something about bees.

2. What topic do you think the writer chose when narrowing the big idea? Why do you think so?

   I think the writer chose bees' food. All the notes tell something about how bees get their food and what they do with their food.

B. Answer the question, based on the above outline.

What kind of organization do you think the writer will use: main idea and details, time order, or compare and contrast? Why do you think so?

Possible response: I think the writer will use main idea and details. The notes don't tell about time order, and they don't compare and contrast.

Writer's Companion ▪ UNIT 5
Lesson 4 *Review Organization*
148

Reaching All Learners

BELOW LEVEL

Help students who are writing below grade level to find the organizing structure through a process of elimination. Remind them that *time order* means steps must be in a certain order. *Compare and contrast* uses comparing words. Then ask what other type of organization they know.

ADVANCED

Have students provide more reasons for Part B. Encourage them to figure out how their reasoning worked as they eliminated the other two possible organizing structures and decided on main idea and details.

ENGLISH LANGUAGE LEARNERS

Encourage students to form groups or pairs that include at least one fairly proficient English speaker. Have them discuss their notes and work together to find the most likely organizing structure.

The Part of an Editied Research Report

OBJECTIVES

- To understand how to edit a research report
- To analyze a Student Model

Teach/Model After reading aloud the introduction, tell students that the writing model on this page is a draft written by a third grader. Read aloud the revising and proofreading tips for how to edit a research report. Then read aloud and review the model. As students listen, have them pay careful attention to the words or sentences that were changed or deleted by the student writer.

Guided Practice Point out the questions below the Student Model. Read aloud the first question. Then guide students in identifying the deletion mark that shows that the word was cut. Briefly review the proofreading marks for deleting, inserting, and capitalization.

Independent Writing Practice Have students complete Items 2–5 independently. Then discuss their responses as a class.

Use a Student Model

LESSON 4: REVIEW ORGANIZATION

Name ______________________

Focus on the Writing Form

The Parts of an Edited Research Report

You have learned how to choose a topic, take careful notes, make an outline, and write a first draft. After you write your draft, you will **edit** it. Here are some tips to help you as you edit.

EDITING TIPS	
REVISE	**PROOFREAD**
Make sure ideas are in an order that makes sense. Make sure all ideas are facts that tell more about the topic. Add words to connect ideas. Combine short, choppy sentences.	Indent paragraphs. Capitalize proper nouns. Make sure subjects and verbs agree. Check spelling and punctuation.

Here is a paragraph from a draft written by a third grader. Notice how the student made marks while revising and proofreading. Then answer the questions.

Student Model

DRAFT

The Anasazi Community
by Keisha

Revised paragraph

Home life (to the Anasazi) was ~~very,~~ very important. Men and women built their homes in the Cliffs and on (top) of mesas. the tops ~~I wouldn't want to live up so high.~~ women and childern took care of the households. ~~Women and children~~ They also made clothing, pots, and many other items. useful

The Anasazi Community
by Keisha

Proofread paragraph

Home life was very important to the Anasazi. Men and women built their homes in the Cliffs and on the tops of mesas. women and childern took care of the households. They also made clothing, pots, and many other useful items.

Use a separate sheet of paper to answer the questions. Revised paragraph

1. Which word was cut out of the first sentence? very
2. Which sentence was cut out of the paragraph completely? I wouldn't want to live up so high.
3. Which phrase in the paragraph was replaced with one word to make the sentences smoother and more interesting? What is the word that was put in place of the phrase? Women and children They

Proofread paragraph

4. Which word was proofread to show it should be a capital letter? Women
5. In which word was a capital letter changed to lower case? Cliffs

Writer's Grammar
When you proofread, put a slash through a capital letter to change it to a lower case letter. Find a slash in the Student Model.

149 Writer's Companion ▪ UNIT 5
Lesson 4 *Review Organization*

WRITER'S GRAMMAR: PROOFREADING MARKS

Read aloud the Editing Tips. Guide students to see the line in the Student Model in which the word *cliffs* is put into lower case. Point out that this is accomplished by puting a slash through the capital letter and making it into a small, or lower case letter. Go on to explain that putting three horizontal lines beneath a lower case letter will turn it into a capital letter. Explain that the delete symbol with its loop and the carat for inserting are common proofreading marks used in editing.

Evaluate an Edited Research Report

OBJECTIVES

- To understand how to evaluate an edited research report
- To see how a 4-point rubric is used to evaluate an edited research report
- To find out why one Student Model received a score of 4 (excellent)

Teach/Model Tell students that, on the next two pages, they will learn how to use a *rubric*. A rubric is a scoring chart that is used to evaluate writing. Turn to page 151 and point out the rubric.

Guided Practice Read aloud the directions for Part A. Then read aloud the Student Model (as edited). Have students listen (without looking at the report) and think about why the Student Model is a success. Then have students read the model and the teacher's comments that go with it. Explain that the comments show some of the things a teacher might look for when evaluating an edited research report. Ask students to identify the comments that have to do with *revising* and those that have to do with *proofreading*.

LESSON 4: REVIEW ORGANIZATION

Name ______________________

Evaluating the Student Model

Evaluate an Edited Research Report

A. **Two students were asked to edit a research report. This report received a score of 4. A score of 4 on a 4-point rubric means "excellent." Read the research report and the teacher comments that go with it. Find out why this research report is a success.**

Student Model

DRAFT

The American Flag
by Betsy

The United States became a country in 1776 ~~. There was~~ and a new flag was needed for the new country. ~~a new flag needed for the new country.~~ Some say that Betsy Ross made the first flag. They say that Betsy Ross was a clothing maker who made clothes for George Washington. ~~Betsy Ross had the same name as I do!~~

There were thirteen colonies in 1776. The flag was made with one star for each of the thirteen colonies. It would be possible to sew on a new star for each new colony. The flag also had thirteen stripes of red and white. That would not change. It would always show how many colonies there were in the beginning.

It is not known for sure if this story is true. It is known that the first flag had thirteen stars and thirteen stripes. It was a new flag for a new country.

Strong work! These sentences are better as a combined sentence.

Glad to see you deleted this sentence. It did not tell an important fact about your topic.

Nice job. You needed to add the word "new" to make this sentence clearer.

Good proofreading. This is the first word of the sentence, so it should be capitalized.

Good proofreading again. This is not a proper noun, so it should not be capitalized.

Writer's Companion ■ UNIT 5
Lesson 4 *Review Organization* 150

WRITER'S STRATEGY: COLD READING

Tell students that one way to double check their work is to go back over it after some time has passed. Guide them to see that this strategy is only possible when time allows, but that it can be very effective. Explain that, when a little time has passed, it is easier to look at the writing as if they are the reader and not the writer. In this way, mistakes can be revealed and corrected.

Evaluate an Edited Research Report

OBJECTIVES

- To read a Student Model that received a score of 2 (needs improvement)
- To understand what makes an edited research report get a score of 2 on a rubric
- To evaluate an essay using a rubric

Teach/Model After reading aloud the directions for Part B, tell students that they will now review a student's edited research report that received a score of 2 on a 4-point rubric. Read aloud the Student Model (as edited). Have students listen (without looking at the words) and think about why the editing of this report needs improvement.

Guided Practice Ask students to comment on any strengths and weaknesses they noticed while listening. Then have students read the Student Model and the teacher's comments that go with it. Guide students in expressing their opinions about why the editing of the report needs improvement.

Independent Writing Practice After reading aloud the directions for Part C, review how to use the 4-point rubric at the bottom of the page. Have students complete it independently.

Use a Student Model

LESSON 4: REVIEW ORGANIZATION

Name ______________________

Evaluating the Student Model

B. This paragraph got a score of 2. Why did it get a low score?

DRAFT

Student Model

by Ben

Our sun is really a star. ~~We have a sun.~~ It is very hot. It is really close to Earth. the sun is much bigger than Earth

The sun gives us heat. It can make snow melt. I like to study the sun more than I like to study planets. What would Earth do without this special star?

Don't forget to give your report a title.

Good job. This is more interesting

Nice job with correcting punctuation.

Good proofreading.

Watch for end punctuation.

Cut this sentence. It is an opinion not a fact.

C. What score would you give the student's story? Put a number on each line.

	4	3	2	1
Purpose ______	☐ The topic is clear. The report includes only facts about the topic.	☐ The report topic is clear. The report includes mostly facts about the topic.	☐ The topic is somewhat clear. There is information that does not tell more about the topic.	☐ The topic and details are not clear. The report may contain opinions.
Organization ______	☐ The beginning, middle, and end are clear. Each paragraph has a topic sentence and supporting details.	☐ The beginning, middle, and end are not clear. Most paragraphs include a topic sentence and supporting details.	☐ The report does not have a beginning, middle, and end. Some paragraphs have details that support the topic.	☐ The report includes few details. Details in the report are not organized.
Grammar ______	☐ The writer uses correct grammar and punctuation.	☐ The writer makes a few mistakes in grammar and punctuation.	☐ The writer makes several mistakes in grammar and punctuation.	☐ The writer makes grammar or punctuation mistakes in almost all the sentences.

151 Writer's Companion • UNIT 5 Lesson 4 *Review Organization*

WRITER'S GRAMMAR: FUN WITH ADJECTIVES

Guide students in understanding that finding vivid adjectives can be lots of fun. Point out that when the writer of the Student Model tells us that the sun is "very hot," he is missing an opportunity. Give students other descriptions of the sun. It could be "boiling hot," "ferociously hot," or "fiercely hot." Have students come up with still more possibilities and have volunteers write these adjectives on the board.

Extended Writing/Test Prep

OBJECTIVES

- To apply the craft of organization in a longer piece of writing
- To select a topic for an extended piece of writing and begin the prewriting process
- To become familiar with the kinds of writing prompts that appear on writing assessments

Teach/Model Tell students that the next two pages will give them an opportunity to select a topic and write a research report. Tell them that they will follow the steps of the writing process while doing this assignment. After reading aloud the directions for Part A, explain that they can choose from three possible choices. Go over the choices listed in Part A and have each student select one.

Guided Practice After reading aloud the directions for Part B, review the steps for organizing a research report. Then have students reread the Student Model tips in the previous lessons:

- Notes: Lesson 1, Page 131
- An Outline: Lesson 2, Page 137
- A Draft: Lesson 3, Page 143
- An Edited Research Report: Lesson 4, Page 149

Then have students begin the prewriting process by completing Part B independently.

LESSON 4: REVIEW ORGANIZATION

Name ______________________

Extended Writing/Test Prep

Extended Writing/Test Prep

On the last two pages of this lesson, you will use what you have learned about organization to write a research report.

A. Read the three choices below. Put a star by the writing activity you would like to do.

1. Look back to the paragraph you wrote on page 142. Develop your paragraph into a complete report.
2. Choose one of the other topics you created on page 147. Write a research report about the topic.
3. Choose a new research report topic you would like to write about.

Respond to a Writing Prompt

REMEMBER—YOU SHOULD

- Decide on your topic.
- Take notes from sources about your topic.
- Organize your notes into an outline.
- Write a first draft and then revise it as you try to use correct spelling, capitalization, punctuation, grammar, and sentences.

USE THIS PREWRITING PAGE TO PLAN YOUR COMPOSITION.

B. Use at least three sources for research. Take careful notes. Make an outline. Then write your report. Begin your planning below.

TOPIC: ______________________

THE KIND OF ORGANIZATION I WILL USE: ______________________

____

NAMES OF SOURCES AND NOTES I TOOK FROM EACH ONE: ______________________

____

____

____

Use additional sheets of paper to take your notes.

Writer's Companion • UNIT 5
Lesson 4 *Review Organization*
152

SHARING AND DISCUSSING

Tell students that finding great sources is critical to writing a good research report. Engage students in a discussion of what makes up a good source. Guide them in understanding that some publications and research tools have better reputations than others. Ask what makes them believe something. Discuss what makes a good source—reliability, transparency, accessibility and so forth.

Extended Writing/Test Prep

OBJECTIVES
- To use a graphic organizer to plan a research report
- To use the steps of the writing process to complete a research report independently

Teach/Model Read aloud the directions to Part C and explain to students that they will select and use a graphic organizer to plan their research report. Provide each student with a photocopy of the appropriate graphic organizer. (See the back pages of this guide.) Then model how to complete a graphic organizer on the board.

Guided Practice: **Prewriting** Provide help as needed while students complete their graphic organizers. Remind students that the graphic organizer is a general guide to use as they write; they can add details to their writing later.

Independent Writing Practice: **Draft, Revise, Publish** Have students write their research reports on a separate sheet of paper. After students have finished their first draft, ask them to revise their work. Remind them to think about how effectively they organized their information. Then have students think about how they would like to publish their work.

LESSON 4: REVIEW ORGANIZATION

Name ______________________

Extended Writing/Test Prep

C. Create an outline on this page for your research report. Use the information from your notes. Be sure to use correct outline form.

____

____

____

____

____

____

____

____

____

____

____

____

____

____

____

____

____

____

____

D. Write your research report on other sheets of paper.

153 Writer's Companion • UNIT 5
Lesson 4 *Review Organization*

WRITER'S STRATEGY: SELECTING A GRAPHIC ORGANIZER

Explain to students that different graphic organizers are used for different kinds of writing. Tell students that they should choose the one that best fits what they would like to write. Here are some suggestions.
- A chart with boxes is a great way to convert an outline into a beginning, middle, and end.
- A flow chart can help when organization is based on time order.
- A chart with columns can help in brainstorming details you will need to support main ideas.

Answering Multiple-Choice Questions

OBJECTIVES

- To understand that some writing tests contain multiple-choice questions
- To become familiar with a multiple-choice test format that requires shading in answer bubbles
- To recognize that the words uncle and aunt are only capitalized if they come before a name

Teach/Model Tell students that some writing tests include multiple-choice questions. Explain that this lesson will help them learn strategies for how to answer this type of question. Read aloud the directions for Part A and the test tip.

Guided Practice Ask a volunteer to read aloud Question 1 and then have students read the three answer choices silently to themselves. Ask them to find the sentence that is correct. Have students mark the correct answer next to C. Then review the rules for capitalization as you model finding the correct answer for Question 2. Demonstrate how to mark the answer.

Independent Writing Practice Have students complete the last two items independently. Review the use of capitalization and punctuation if necessary.

LESSON 5: WRITING TEST PRACTICE

Name ______________________________

Writing Test Practice

Answering Multiple-Choice Questions

Some writing tests have questions with answer choices. This lesson will help you practice this kind of test.

A. Some test questions may ask you about sentences. Read the test tip. Then shade the bubble next to the correct answer for each question.

Read and answer questions 1–4.

1. In which sentence below is all **punctuation** correct?
   - Ⓐ He bought carrots, and peas.
   - Ⓑ He, bought carrots and peas.
   - Ⓒ He bought carrots and peas.
2. In which sentence below is all **capitalization** correct?
   - Ⓕ We traveled to another State to visit my Uncle.
   - Ⓖ We traveled to another state to visit my uncle.
   - Ⓗ We traveled to another State to visit my uncle.
3. In which sentence below is all **punctuation** correct?
   - Ⓐ I ran to the tree the lake and then home.
   - Ⓑ I ran to the tree the lake, and then home.
   - Ⓒ I ran to the tree, the lake, and then home.
4. In which sentence below is all **capitalization** correct?
   - Ⓕ Her grandmother lives in Texas.
   - Ⓖ Her grandmother lives in texas.
   - Ⓗ Her Grandmother lives in Texas.

Test Tip: In question 2, only capitalize the words *uncle* and *aunt* if they come before a name. Example: Aunt Tam.

SHARING AND DISCUSSING

Have students compare their responses to the last two items. Suggest they brainstorm non-family words that are capitalized when they appear before names, as with such professional titles as doctor and senator.

Answering Multiple-Choice Questions (cont.)

OBJECTIVES

- To understand that answering some multiple-choice questions involve reading a passage
- To become familiar with reading a passage in which some sentences are numbered
- To recognize that the verb and subject of a sentence must agree

Teach/Model Tell students that some multiple-choice items ask them to read a passage and then answer questions about it. Read aloud the directions for Part B and have students read the passage silently to themselves. Explain that the sentences are numbered because the test items identify the sentences by number. Read aloud the test tip as a reminder to check for subject-verb agreement. Then read aloud the first test item and model looking back at the passage to find the correction in Sentence 1. Model selecting the correct answer and marking the circle on the Answer Sheet.

Guided Practice Guide students in using the same steps to complete the second item.

Independent Writing Practice Have students answer Item 3 independently.

LESSON 5: WRITING TEST PRACTICE

Name ______________________________

Writing Test Practice

B. Some test questions may ask you to read a passage and answer questions about it. The sentences in the passage may be numbered. Read the test tip.

DIRECTIONS

Read the passage below. Then read each question and fill in the correct answer on your Answer Sheet.

The Fair

(1) We likes to go to the fair. (2) We has gone many times. (3) Our fair is the best in the whole city. (4) We try to get there early. (5) We always stays in groups with friends and neighbors for safety.

1. What change, if any, should be made in sentence 1?
   - A Change ***We*** to **Us**
   - B Change ***to*** to **too**
   - C Change ***likes*** to **like**
   - D Make no change

2. What revision, if any, is needed in sentence 2?
   - F We have gone many times.
   - G We has goed many times.
   - H We is going many times.
   - J No revision is needed.

3. What revision, if any, is needed in sentence 5?
   - A We always stay in groups with friends, and neighbors for safety.
   - B We always stay in groups with friends and neighbors for safety.
   - C We always be staying in groups with friends and neighbors for safety.
   - D No revision is needed.

Test Tip: Find the action word in the sentence. This is the verb. Then figure out who is doing the action. This is the subject. Be sure the subject and verb agree.

Answer all test questions on this Answer Sheet.

1. Ⓐ Ⓑ ● Ⓓ
2. ● Ⓖ Ⓗ Ⓙ
3. Ⓐ ● Ⓒ Ⓓ

EXTENDING THE CONCEPT: WORD CHOICE

If a student answers a test item incorrectly, try to find out why. An incorrect response may result from marking an answer in the wrong space on the Answer Sheet. Then review with students how to find and mark the correct answers on both an Answer Sheet and a fill-in test.

Answering Multiple-Choice Questions (cont.)

OBJECTIVES

- To understand that answering some multiple-choice questions involve choosing the best way to revise or correct a sentence
- To become familiar with filling in the blanks in a boxed sentence with choices that appear below it
- To review that words that describe things are called adjectives

Teach/Model Read aloud the directions for Part C and have students read the first sentence silently to themselves. Explain that the two words in each choice fills the two blanks in the sentence. Then read aloud the first test item and model selecting the correct answer and marking it on the Answer Sheet.

Guided Practice Guide students in answering the second test item. Ask volunteers to explain their answers. Then read aloud and review the test tip.

Independent Writing Practice Have students complete the last two items independently.

LESSON 5: WRITING TEST PRACTICE

Name ______________________

Writing Test Practice

C. Some test questions may ask you the best way to revise or correct sentences. Read the test tip.

Read the questions and answer them on the Answer Sheet.

1. Read this sentence.

I want to ________ sure to ________ carrots for a tasty treat.

Which pair of words makes the sentence correct?

A have, cutted
B be, cut
C was, cutted
D am, cut

Test Tip: Some words are used to describe things. These words are called adjectives *(happy, old)*.

2. Which sentence is written correctly?

F My cousins and Uncles are coming tomorrow for Thanksgiving.
G My cousins and uncles are coming tomorrow for Thanksgiving.
H My cousins and Uncles are coming tomorrow for thanksgiving.
J my cousins and uncles are coming tomorrow for Thanksgiving.

3. Which two words can describe something?

A bake, cake
B good, cake
C good, bad
D bake, bad

4. Which two words show action?

F bake, make
G good, cake
H bake, cake
J bake, bad

Answer all test questions on this Answer Sheet.

1. Ⓐ **B** Ⓒ Ⓓ
2. Ⓕ **G** Ⓗ Ⓘ
3. Ⓐ Ⓑ **C** Ⓓ
4. **F** Ⓖ Ⓗ Ⓘ

Writer's Companion ■ UNIT 5
Lesson 5 *Writing Test Practice* 156

Reaching All Learners

BELOW LEVEL

Have students who are performing below grade level restate in their own words what each test item is asking them to do. Emphasize that each of these items requires a different type of task. Make sure that it is clear in the last item that both words in their choice must be describing words.

ADVANCED

Have students brainstorm other describing words, similar to the ones given in the answer choices of these items. Challenge students to branch out and brainstorm *adverbs*, which are describing words that modify verbs instead of nouns.

ENGLISH LANGUAGE LEARNERS

Have students work in groups to review what they know about action words and tenses as they go over their answers to Question 1. Explain that both action words that fill in the blanks must be in the same tense—which is present tense.

Answering Multiple-Choice Questions (cont.)

OBJECTIVES

- To understand that answering some multiple-choice questions involves reading a passage
- To become familiar with reading a passage in which numbered blanks appear
- To remember to only capitalize place names when they refer to a specific place

Teach/Model Tell students that some multiple-choice items ask them to read a passage and then answer questions. Read aloud the directions for Part D and have students read the passage silently to themselves. Explain that the blanks are numbered because the test items identify them by number and ask for the correct word to fill in each blank. Read aloud the test tip. Then read aloud the first test item and model looking back at the passage to find blank (1) in the first sentence. Model selecting the correct answer and marking the circle next to it.

Guided Practice Guide students in using the same steps for Items 2–3.

Independent Writing Practice Have students answer Item 4 independently.

LESSON 5: WRITING TEST PRACTICE

Name ______________________

Writing Test Practice

D. **Some test questions may ask you about sentences in a paragraph. Read the test tip. Then read the passage and each question and mark the circle next to the correct answer.**

Read *Party Day*. Choose the words that correctly complete questions 1–4.

Party Day

Today was party day at our __(1)__. We got ready. We cleared away our books and papers. We __(2)__ done this for parties before. Family members started to come. They brought juice and games. Nita's aunt came to help. I __(3)__ my grandfather come in next. Today's party __(4)__ better than our party last year.

1. Which answer should go in blank (1)?
   - Ⓐ school's
   - Ⓑ School
   - Ⓒ school
2. Which answer should go in blank (2)?
   - Ⓕ had
   - Ⓖ was
   - Ⓗ has
3. Which answer should go in blank (3)?
   - Ⓐ see
   - Ⓑ seen
   - Ⓒ saw
4. Which answer should go in blank (4)?
   - Ⓕ will be
   - Ⓖ was
   - Ⓗ is

Test Tip: Remember that the name of a place is only capitalized when you are naming a specific place. In question 1, the name of the school is not given.

157 **Writer's Companion ▪ UNIT 5**
Lesson 5 *Writing Test Practice*

USING ACADEMIC LANGUAGE

Most tests contain academic language. Academic language includes the names of grammar words such as *subject, verb, adjective,* and *capitalize.* Preparing students to do well on tests includes helping them recognize and understand academic language. Before asking students to complete the above test items, discuss the meaning of these terms: *present tense* and *past tense.* Ask students how they determined which verb was correct for each of the last two items on this page. Explain that often something just sounds wrong to our ear, and that the rules of grammar can often explain *why* it sounds wrong.

Look at Imagery and Figurative Language

OBJECTIVES

- To understand the terms *imagery* and *figurative language*
- To identify imagery and figurative language in literature

Teach/Model Read aloud the introduction. Explain that imagery uses images, or pictures, to help readers see what the writer is describing. Tell students that figurative language refers to words and phrases that create imagery. Read aloud the poem from *I'm in Charge of Celebrations*. Have students listen for imagery and figurative language.

Guided Practice Read aloud the directions for Part B. For Item 1, guide students in finding the figurative language that compares one idea to another. Then discuss why the language is figurative. Ask how the language makes something happen that is unexpected, unusual, or impossible. Guide students in completing Part B.

Independent Writing Practice Have a volunteer read aloud the directions for Part C. Then have students complete it independently by giving a word that is used to compare ideas or things.

Use a Literature Model

LESSON 1: IMAGERY AND FIGURATIVE LANGUAGE

Name ______________________________

Writer's Craft in Literature

Look at Imagery and Figurative Language

Imagery is description that helps your readers make pictures in their minds. **Figurative language** is a kind of imagery. It can be used to compare one thing to another.

A. Read the Literature Model. Think about how the writer uses imagery in the stanzas of the poem.

Literature Model

You can tell
what's worth
a celebration
because
your heart will
POUND
and
you'll feel
like you're standing
on top of a mountain
and you'll
catch your breath
like you were
breathing
some new kind of air.

You know how
they come
from far away,
moving
up from the flats,
swirling
and swaying
and falling
and turning,
picking up sticks
and sand
and feather
and dry tumbleweeds.

—from *I'm in Charge of Celebrations*
by Byrd Baylor

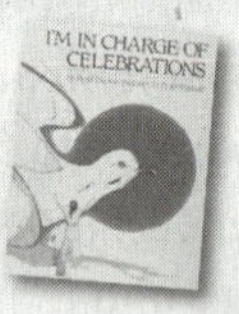

B. Identify imagery and figurative language.

1. Circle examples of figurative language that compare.
2. Underline examples of imagery that describe without comparing. Answers may vary.

C. Read the figurative language in the poem. Which word helps you know that the writer is comparing? Write your answer. like

Writer's Companion ■ UNIT 6
Lesson 1 *Imagery and Figurative Language* 158

SHARING AND DISCUSSING

Explain that people use comparisons all the time to help people get a better picture of what is meant. Offer a comparison that uses figurative language, such as *It was hot enough to fry an egg on the sidewalk.* Invite students to provide other examples that complete a sentence, such as *It was so cold that...* [students complete the sentence].

Explore Imagery and Figurative Language

OBJECTIVES

- To deepen understanding of imagery and figurative language
- To identify and explore the use of sensory words

Teach/Model Use the graphic organizer at the top of the page to help students understand how writers use sensory words. Remind them that these are words that involve the five senses and help the reader imagine what the writer is trying to communicate. Explain that writers use sensory language to make a strong impression on the reader. If the reader can imagine seeing, hearing, tasting, feeling, or smelling an idea, he or she has a better idea of what the writer wants to say.

Guided Practice Read the directions for Part A aloud and model using the information provided in the example to find the sensory word or phrase. Guide students in completing Part A.

Independent Writing Practice After reading aloud the directions, have students complete Part B independently. For Part C, see Extending the Concept.

LESSON 1: IMAGERY AND FIGURATIVE LANGUAGE

Name ______________________

A Closer Look at Writer's Craft

Explore Imagery and Figurative Language

Writers use sensory words to paint strong word pictures, or imagery.

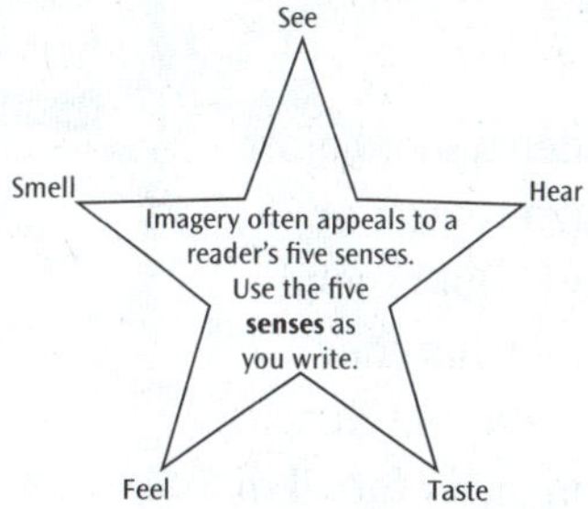

A. Read the sentences. Answer the questions to tell how the writer paints word pictures.

Example I touched the soft teddy bear.
Underline the word the writer uses to tell how the teddy bear feels.

I woke up and smelled sweet muffins baking. Then I raced outside and grabbed the icy, hard snow. Icicles hung from the roof like long, clear, pointed fingernails reaching to make snowballs.

1. Underline the word that tells how the muffins smelled.
2. Circle the words that tell how the snow felt.
3. Box the words that tell how the icicles looked.

B. Read this stanza from *I'm in Charge of Celebrations*. Underline the names of colors used to describe the cloud.

It was not
bluish-green
or grayish-green
or something else.

This cloud
was
green . . .
green as a jungle parrot.

C. Circle the line that shows the comparison used to describe the cloud.

EXTENDING THE CONCEPT: DESCRIBING CLOUDS

Before students complete Part C, have them think about the way in which the cloud has been described in the poem. How is this comparison like, or unlike, other ways in which they have heard clouds described? Tell students to write down a phrase that they might use to describe a cloud or clouds.

Use Imagery and Figurative Language

OBJECTIVES

- To prepare to write a poem by choosing a topic
- To practice writing sensory words that create imagery

Teach/Model Tell students that before they write a poem about a season or an outdoor activity, they should spend some time thinking about what they want to write about. Explain that they do not have to include imagery for all of the senses in their poems. Read aloud the introduction and the example then discuss the senses the writer uses in the unrhymed poem.

Guided Practice Read aloud the directions for Part A. Then have students write the name of a season or something they do outdoors. Guide them in filling out the star with imagery that has to do with each sense.

Independent Writing Practice Have students complete Part B independently. Invite them to share their stanzas aloud.

LESSON 1: IMAGERY AND FIGURATIVE LANGUAGE

Name ______________________

Practice with Writer's Craft

Use Imagery and Figurative Language

Before you write an unrhymed poem, you may want to jot down some imagery to include. Here is how one third grader prepared to write a poem about summer. Notice that he chose to appeal to only two senses.

Example Title: *Summer Fun*

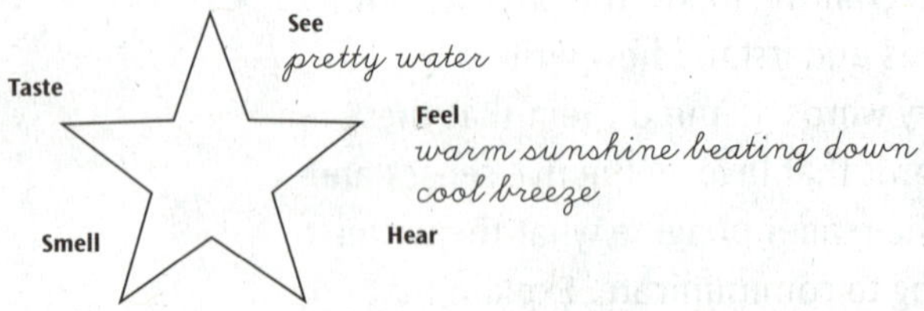

A. Prepare to write an unrhymed poem about a season or something you do outdoors. Write the title. Then fill in your star. Try to find some imagery for each sense, if possible. Answers will vary, but students should provide phrases that match senses.

Title: ______________________

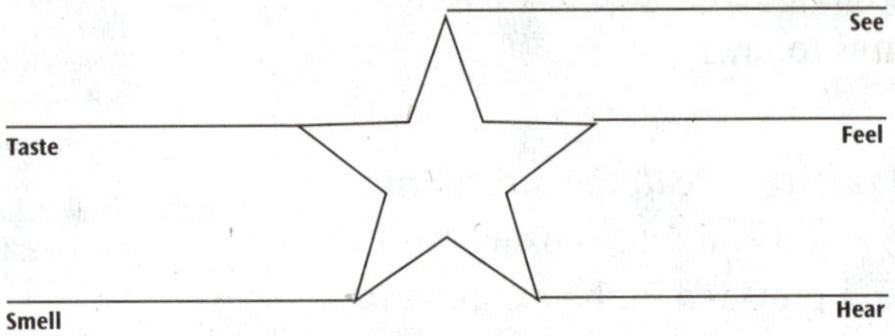

B. Use some of the senses from your star. Write the first stanza of your poem on another sheet of paper. Answers will vary, but the answer should come from the student's star.

Writer's Companion ▪ UNIT 6
Lesson 1 *Imagery and Figurative Language* 160

Reaching All Learners

BELOW LEVEL

Pair each student who is writing below grade level with an aide or a classmate. Tell the student to orally describe the season or outdoors experience he or she wants to write about. Ask the aide or partner to help the student write down details that include sensory words.

ADVANCED

Encourage students writing above grade level to think about which season they would want to experience for most of the year, and why. Have them try to use some of this information into their poems.

ENGLISH LANGUAGE LEARNERS

Pair less fluent English speakers with more fluent English speakers. Have the more fluent English speakers help their partner identify appropriate words for the sensory words their partner describes.

The Parts of a Poem

OBJECTIVES
- To understand how a poem is developed
- To analyze a Student Model

Teach/Model Read aloud the introduction. Explain to students that the writing model on this page is a first draft of a student's poem. Read aloud the steps for how to develop a poem. Explain that the words at the end of the lines do not rhyme in this poem. Then read the model aloud. As students listen, have them pay careful attention to how the poem is developed with images.

Guided Practice Point out the items below the Student Model. Read the first item aloud. Then guide students in finding the title of the poem and drawing an arrow to it. Read the second item aloud and guide students in underlining the word that describes sunshine

Independent Writing Practice Have students complete Items 3–5 independently. Then discuss how the poem uses words to convey images to the reader.

Use a Student Model

LESSON 1: IMAGERY AND FIGURATIVE LANGUAGE

Name ______________________

Focus on the Writing Form

The Parts of a Poem

A good **poem** guides readers to paint pictures in their minds. You can read a student's first draft of a poem below. Think about how the writer uses imagery to guide readers. Then answer the questions.

Student Model

DRAFT

Summer Fun ←
by Chase

The warm sunshine
beating down
feels so nice.

It follows me
to the pool
where I jump
into the pretty water.
The water feels like
a cool breeze.

Write a title for the poem.

Introduce the topic and write down the first image.

Add descriptive words and figurative language to describe the image.

Write about another image and repeat step two.

Check that the lines break in places that will help readers understand the poem.

1. Draw an arrow to the title of the poem.
2. Underline a word that describes the sunshine.
3. Circle the sentence that compares.
4. What does the writer describe first? sunshine
5. What does the writer describe next? water in the pool

Writer's Grammar
Poems are divided into stanzas. A line is usually skipped between stanzas. Count the number of stanzas in the Student Model.

161 Writer's Companion • UNIT 6
Lesson 1 *Imagery and Figurative Language*

WRITER'S GRAMMAR: STANZAS

Have students draw a box around each section of the Student Model. Explain that these sections are called *stanzas*. Explain that when they create their poems, they should skip a line between stanzas.

Evaluate a Poem

OBJECTIVES

- To understand what to look for when evaluating a poem
- To evaluate a poem using a checklist
- To write an evaluation of a poem

Teach/Model Discuss with students why it is important to evaluate writing. Explain that evaluating what they have written helps them see how to improve their writing. Discuss what to look for in evaluating a poem, using the points at the top of the page.

Guided Practice Read aloud the directions for Part A and its questions. Guide students in referring to the Student Model on page 161 for information. Tell them that using a checklist will also help them focus on what they should evaluate. Point out the checklist in Part B and explain that it contains some of the writing skills practiced in this lesson

Independent Writing Practice After reading aloud both sets of directions, have students complete Parts B and C independently. Point out the callouts on page 161 and explain that students can use them to help evaluate the poem.

Name ______________________________

Evaluate a Poem

How do you decide if someone has done a good job writing a poem? Ask yourself these questions:

- Does the writer use strong, colorful words that clearly express ideas? (Look for sense words and words that paint strong pictures.)
- Does the poem make interesting comparisons? (Look for words such as *like* or *as*.)

A. Reread the Student Model on page 161. Then answer these questions.

1. Why does the writer describe the sun by saying that it *follows me?*

   Possible response: to paint a strong picture of the warmth in the reader's mind

2. With what does the writer compare the water in the pool?

   The writer compares the water to a cool breeze.

B. Now evaluate the Student Model. Put a check in the box next to each thing the writer did well. Do not check the box if you do not think the writer did a good job.

- ☐ The title and opening lines show the topic of the poem.
- ☐ The poem has images that show how the writer feels about the topic.
- ☐ The writer used strong, colorful words to describe the images.
- ☐ The poem makes interesting comparisons.
- ☐ The poem has sense words that paint pictures in the reader's mind.

C. How do you think the writer could make the poem better? Write your ideas on another sheet of paper. Possible response: The writer could give clearer images. The writer could make more comparisons.

See the rubric on page 207 for another way to evaluate the Student Model.

EXTENDING THE CONCEPT: USING IMAGES

Explain that in the second stanza of the Student Model, the writer has created an interesting image—that of sunshine following the narrator. Tell students that the writer is introducing an unusual image, because people know that sunshine does not really follow one specific person. Ask them for additional ideas of what the sunshine could do that would be unusual or interesting to help create a strong image for the reader.

Revise by Creating a Strong Image

OBJECTIVES

- To revise a poem by creating a strong image
- To revise writing by using sensory words and figurative language

Teach/Model Tell students that revising their writing gives them a chance to make changes that improve their work. Explain that in this activity, they will revise by changing, adding, or deleting words and phrases to help them create a strong image.

Guided Practice Read the introduction and the example aloud. Help students understand that the comparison to a hug helps the reader get a much better picture of the warm, cozy effect of the sun. Ask a volunteer for a different comparison that might be made about a sun beating down. Then read aloud the directions for Part A. Guide students in completing Item 1 by finding other choices that create a better word picture.

Independent Writing Practice Have students complete Items 2–3 of Part A on their own. Then read aloud the directions for Part B and have students complete it independently. Invite volunteers to share their revisions aloud.

LESSON 1: IMAGERY AND FIGURATIVE LANGUAGE

Name ______________________

Revising the Student Model

Revise by Creating a Strong Image

What is one thing the student writer could have done better? The writer could have created stronger images. Here is an example of how a sentence from the Student Model can be improved.

Example

The warm sunshine
beating down
feels so nice.

The warm sunshine
beating down
feels like a welcoming hug.

A. **Revise these parts of the Student Model. Make the word pictures stronger. Add new word pictures.** Possible responses given

1. It follows me *It tiptoes after me*

   to the pool to the shining pool

   where I jump where I leap joyfully

   into the pretty water. into the dancing water.

2. Write a new sentence to add to the first stanza. Create strong word pictures with your sentence.

   I can feel the heat pour through to my toes.

3. Write a new sentence to add to the second stanza. Create strong word pictures with your sentence.

   Then I come back out into the fire of day.

B. **Revise the stanza you wrote on page 160. Add figurative language and other imagery. Use another sheet of paper.**
Answers will vary, but should include strong sensory images and comparisons.

163 Writer's Companion ▪ UNIT 6
Lesson 1 *Imagery and Figurative Language*

WRITER'S STRATEGY: WORDS THAT SING

Tell students that after they have completed revising their poems, they should share them with a partner. Have them ask, "Do the words *sing?*"—meaning, do they convey strong images? Do the words paint a picture that makes the poem come alive for the reader? After they have listened to their partners' reactions, students should review their poems once again to make sure they have used words that are exciting and that "sing."

Look at Vivid Verbs

OBJECTIVES

- To understand the term *vivid verbs*
- To identify vivid verbs in literature

Teach/Model Read aloud the introduction and explain that vivid verbs make reading more exciting. Tell students that vivid verbs bring action to life with strong and clear words. Read aloud the passage from *Alejandro's Gift*. Have students listen for vivid verbs.

Guided Practice Read aloud the directions for Part B and Item 1. Guide students in finding all of the verbs in the passage. Then discuss why some verbs are more interesting than others. Ask what is the difference in how readers might react to reading *scampered* instead of *ran*. Guide students in completing the rest of Part B.

Independent Writing Practice Have students complete Part C independently by giving examples of verbs that could replace *looked*.

Use a Literature Model

LESSON 2: VIVID VERBS

Name ______________________________

Writer's Craft in Literature

Look at Vivid Verbs

Vivid verbs are action words that give strong, clear descriptions of what is happening. Vivid verbs help make language more colorful and interesting.

A. Read the following Literature Model. Think about how the writer uses vivid verbs.

Literature Model

A ground squirrel crept from the underbrush. Moving warily over the sand, it hesitated and looked around. Alejandro paused, keeping very quiet as the squirrel approached the garden. It ran up to one of the furrows, drank its fill of water, and scampered away. After it left, Alejandro realized that for those few moments his loneliness had been all but forgotten.

—from *Alejandro's Gift*
by Richard E. Albert

B. Identify vivid verbs. Answers may vary.

1. Underline the verbs in the passage.
2. Circle vivid verbs that tell how the animals moved.

C. Write two vivid verbs that could replace *looked* in the paragraph.

Possible responses: glanced, gazed

Writer's Companion ▪ UNIT 6
Lesson 2 *Vivid Verbs*

164

EXTENDING THE CONCEPT: ACTING OUT VIVID VERBS

Help students visualize the concept of vivid verbs by giving them some examples and having them act out the words. Divide students into small groups. Give each student in the group a card with a vivid verb, such as *skipped, leaped, hopped,* or *jumped*. Have students act out each word for the other members of the group. Group members should guess which verb the student is acting out.

Explore Vivid Verbs

OBJECTIVES

- To deepen understanding of vivid verbs
- To identify and explore the difference between vivid verbs and unclear verbs

Teach/Model Use the graphic organizer at the top of the page to help students understand that some verbs paint a stronger and clearer picture for the reader than others do. Give an example of an unclear or weak verb, such as ran, and a stronger choice, such as *dashed* or *hurried.*

Guided Practice Read the directions for Part A aloud and model using the Word Box to find a vivid verb that could replace the verb in the example. Guide students in completing Part A.

Independent Writing Practice After a volunteer reads aloud the directions, have students complete Part B independently. For Part C, see Sharing and Discussing.

Name ______________________

A Closer Look at Writer's Craft

Explore Vivid Verbs

Good writers use vivid verbs to make their writing interesting and clear. They form pictures in their own minds as they write.

Unclear Verb
gives a big idea
does not paint a clear picture
does not tell exactly what the writer means
Examples: ran, looked, moved

Clear, Vivid Verb
paints a clear picture
tells exactly what the writer means
Examples: slithered, galloped, glared

A. **Read each sentence. Rewrite the sentence by replacing the underlined verb with a vivid verb. If you wish, use verbs from the Word Bank.** Possible responses given.

Example He asked for just one more chance.

He begged for just one more chance.

1. The bear went away into the forest. dashed, crept
2. She said that she could do the job even better. boasted, shouted
3. One puppy took the bone away from the other. grabbed, stole

B. **Read these sentences from *Alejandro's Gift*. Circle the one that gives a clearer description by using vivid verbs.**

The squirrel did come again, from time to time bringing along small friends.

These were times he cherished, and he often stayed for hours, working until driven indoors by the desert heat.

Word Bank
dashed
boasted
crept
grabbed
stole
shouted

C. **Rewrite the sentence you did NOT circle above. Replace the unclear verbs with clear, vivid verbs. Use another sheet of paper.**
Possible response: The squirrel did race back again, from time to time, trailed by small friends.

SHARING AND DISCUSSING

Before students complete Part C, have them think about the squirrel's movement. Ask: *What might he be doing, and in what way is he acting?* As a class, brainstorm ways in which the squirrel could be moving. On the board, write a list of vivid verbs, such as *scurry, dart,* or *rush,* that students could use in their rewritten sentences. Tell them to rewrite as much of the rest of the sentence as possible.

Use Vivid Verbs

OBJECTIVES

- To prepare to write a thank-you letter by choosing a topic
- To practice writing vivid verbs

Teach/Model Tell students that before they write a thank-you letter for a gift, or a favor, they should spend some time thinking about what they want to say. Explain that by using vivid verbs, they will make their letters strong and clear. Read aloud the introduction and the example. Then discuss why the verbs listed by the writer are good choices.

Guided Practice Read aloud the directions for Part A. Then have students write two verbs they plan to use, as well as other choices that are more vivid. Guide them in filling out the chart.

Independent Writing Practice Have students complete Part B independently. Invite them to share some of their vivid verbs and explain why they chose them.

LESSON 2: VIVID VERBS

Name ______________________

Practice with Writer's Craft

Use Vivid Verbs

Prepare to write a thank-you letter. Think about why you are writing the letter and who will receive it. Will your verbs be clear and strong? Here is how one student made a list of vivid verbs while preparing to write a letter.

Example **Vivid Verbs**

Instead of *went:*	Instead of *fell:*
dashed, leaped, rushed, crawled, hopped, skipped, ran, drove	crashed, floated, tumbled

A. Think about verbs you plan to use in your letter. Write two of the verbs on the lines below. Then write vivid verbs to replace each one. Answers will vary.

Vivid Verbs

Instead of ____________:	Instead of ____________:

B. Choose verbs from your chart that make your meaning strong and clear. Use them to write a thank-you letter for a gift or a favor. Use another sheet of paper for your writing. Answers will vary, but paragraph should include vivid verbs from chart. Vivid verbs should be circled.

Writer's Companion ■ UNIT 6
Lesson 2 *Vivid Verbs*

166

Reaching All Learners

BELOW LEVEL

Pair students who are writing below grade level with an aide or a classmate. Tell the student to orally describe whom he or she wants to thank and the reason for the letter—a gift or a favor. Ask the aide or partner to help the student write down details that include vivid verbs.

ADVANCED

Encourage students writing above grade level to write a longer and more detailed thank-you letter. Encourage them to tell what the gift or favor meant to them. Remind them to include vivid verbs.

ENGLISH LANGUAGE LEARNERS

Pair less fluent English speakers with more fluent English speakers. Have the more fluent English speakers supply appropriate words for vivid verbs their partner describes. Have the more fluent English speakers explain the standard parts of a thank-you letter to their partners.

The Parts of a Thank-You Letter

OBJECTIVES
- To understand how a thank-you letter is organized
- To analyze a Student Model

Teach/Model Explain to students that the writing model on this page is a first draft of a student's thank-you letter. Read aloud the steps for how to organize a thank-you letter. Explain that letters have a heading, a greeting, and a closing. Then read the model aloud. As students listen, have them pay careful attention to the vivid verbs used in the thank-you letter.

Guided Practice Point out the items below the Student Model. Read the first item aloud. Then guide students in finding the heading of the thank-you letter and underlining it. Read the second item aloud and guide students in drawing an arrow to the greeting.

Independent Writing Practice Have students complete Items 3–5 independently. Then discuss how vivid verbs strengthen the impact of the writer's thank-you letter.

Use a Student Model

LESSON 2: VIVID VERBS

Name ____________________

Focus on the Writing Form

The Parts of a Thank-You Letter

A good **thank-you letter** uses vivid verbs to explain why the writer is offering thanks. As you read the letter below, think about how the student writer uses vivid verbs. Then answer the questions.

Student Model

DRAFT

12 Maple Street
Orlando, Florida 32805
April 10, 20--

Write the address, date, and greeting.

Dear Aunt Sophie,

Thank you so much for taking me camping. After we walked through the trees, I liked seeing the beautiful lake. The water bubbled over the rough rocks and went down the side of the cliff.

I got a little bit scared when the tent flapped in the wind at night. You made me feel better, though. You said that everything would be okay. I hope we will be able to go back to the campground soon. Thank you again.

Begin the body by stating what you are thanking the person for. Restate the thank you at the end.

Provide supporting examples. Include vivid verbs and specific nouns to add detail.

Conclude the letter.

Love,
Riga

Write the closing, and sign the letter.

1. Underline the heading.
2. Draw an arrow to the greeting.
3. Circle the body of the letter.
4. Write two vivid verbs from the letter. bubbled, flapped
5. Why are the vivid verbs important in the letter?

   They make the description clear and interesting.

Writer's Grammar
Place a comma after the closing in your letter. Write your signature on the line below the closing.

167 Writer's Companion ■ UNIT 6
Lesson 2 *Vivid Verbs*

WRITER'S GRAMMAR: THE CORRECT FORM OF A THANK-YOU LETTER

Tell students that all letters have a *standard,* or correct, form. For a thank-you letter, explain that they should place a comma after the closing and write their signature on the line below the closing.

Evaluate a Thank-You Letter

OBJECTIVES

- To understand what to look for when evaluating a thank-you letter
- To evaluate a thank-you letter using a checklist
- To write an evaluation of a thank-you letter

Teach/Model Discuss with students why it is important to evaluate writing. Explain that evaluating what they have written helps them find ways to improve their writing. Discuss what to look for in evaluating a thank-you letter, using the points at the top of the page.

Guided Practice Read the directions for Part A and its questions aloud. Guide students in referring to the Student Model on page 167 for information. Tell them that using a checklist will help them focus on what they should evaluate. Point out the checklist in Part B and explain that it contains the writing skills practiced in this lesson.

Independent Writing Practice After reading aloud both sets of directions, have students complete Parts B and C independently. Point out the callouts on page 167 and explain that students can use them to evaluate the thank-you letter.

LESSON 2: VIVID VERBS

Name ______________________

Evaluating the Student Model

Evaluate a Thank-You Letter

When you evaluate a thank-you letter, ask yourself these questions:

- Does the letter include all the parts: heading, greeting, body, closing, and signature? (Look for the address and date, the *Dear* line, the paragraphs, and the signature below the closing.)
- Does the letter use vivid verbs and tell why she is thanking someone? (Look for verbs that are clear and strong.)

A. Reread the Student Model on page 167. Then answer these questions.

1. Why is the student writing to thank her aunt?
   for taking her on a camping trip
2. What vivid verbs helped you picture the camping trip?
   Possible responses: bubbling water, tent flapping in the wind

B. Now evaluate the Student Model. Put a check in the box next to each thing the writer did well. Do not check the box if you do not think the writer did a good job.

- ☐ The letter has an address, date, and greeting.
- ☐ The letter includes a body that states what the writer is offering thanks for.
- ☐ The writer used many vivid verbs in supporting details.
- ☐ The letter includes a closing.
- ☐ The writer signed the letter.

C. How do you think the writer could make the letter better? Write your ideas below.

Possible response: The writer could replace the unclear verbs with vivid verbs.

____

Writer's Companion ▪ UNIT 6
Lesson 2 *Vivid Verbs* 168

See the rubric on page 207 for another way to evaluate the Student Model.

SHARING AND DISCUSSING

Invite students to share their reactions to the Student Model. Ask them to think about how Aunt Sophie might have felt when she received it. Discuss what are the strongest ideas in the letter and what pictures the writer painted for the reader.

Revise by Adding Vivid and Specific Words

OBJECTIVES

- To revise a thank-you letter by adding vivid and specific words
- To revise writing by using words that strengthen its impact

Teach/Model Tell students that revising their writing gives them a chance to make changes that improve their work. Explain that in this activity students will revise by adding vivid and specific words.

Guided Practice Read the introduction and the example aloud. Help students understand that the word *went* is not a strong, clear, or vivid verb. Share the image that the word *raced* conveys. Read aloud the directions for Part A. Guide students in completing Item 1 by finding other verb choices that are more vivid.

Independent Writing Practice Have students complete Items 2–3 of Part A and Part B independently. Invite volunteers to share their revisions aloud.

Name ______________________

Revising the Student Model

Revise by Adding Vivid and Specific Words

What is one thing the student writer could have done better? The writer could have added more vivid verbs and other specific words. Here is an example of how a sentence from the Student Model can be improved.

Example The water bubbled over the rough rocks and went down the side of the cliff.

The water bubbled over the rough rocks and raced down the side of the cliff.

A. **Revise these parts of the Student Model. Replace each underlined word to make it more vivid and clear.** Possible responses are given.

1. After we walked through the trees,

   After we tromped through the trees,

2. I liked seeing the beautiful lake.

   I liked seeing the deep blue lake.

3. You said that everything would be okay.

   You whispered that everything would be okay.

B. **Revise the thank-you letter you wrote on page 166. Add vivid verbs and other vivid, clear words. Finish your writing on another sheet of paper.**

Answers will vary, but should include vivid, specific words.

WRITER'S STRATEGY: VERBS THAT SHOW ACTION OR FEELING

Tell students that after they have completed revising their thank-you letters, they should review their thank-you letters once more to make sure that they have used verbs that are strong. Explain that students should reread all the verbs. Ask them whether their verbs convey enough action or feeling. Is there another, better choice? Encourage them to use a dictionary or thesaurus during the revision process.

Look at Vivid Verbs and Specific Nouns

OBJECTIVES

- To understand the terms *vivid verbs* and *specific nouns*
- To identify vivid verbs and specific nouns in literature

Teach/Model Explain that vivid verbs and specific nouns make the writer's meaning clearer to the reader. Tell students that vivid verbs show actions, and specific nouns give precise names to people, places, or things. Read aloud the passage from *Rocking and Rolling*. Have students listen for vivid verbs and specific nouns.

Guided Practice Read aloud the directions for Part B. For Item 1, guide students in finding the vivid verbs in the passage. Then discuss why a verb might be termed *vivid*. Guide them in completing the rest of Part B.

Independent Writing Practice Have students complete Part C independently by giving an example of a verb from the passage that shows the sound of the wind.

Use a Literature Model

LESSON 3: VIVID VERBS AND SPECIFIC NOUNS

Name ______

Writer's Craft in Literature

Look at Vivid Verbs and Specific Nouns

Good writers use **vivid verbs** and **specific nouns**. These action and naming words clearly show the writer's meaning.

A. **Read the following Literature Model. Think about how the writer uses vivid verbs and specific nouns.**

Literature Model

The answer is blowing in the wind. Day after day it whistles through the valley, picking up grit and sand and blasting everything it touches.

The wind works like sandpaper, slowly wearing the rocks down and grinding them into weird and wonderful shapes.

—from *Rocking and Rolling* by Philip Steele

B. **Identify vivid verbs and specific nouns.** Answers may vary.

1. Underline the vivid verbs in the paragraph.
2. Circle the specific nouns in the paragraph.

C. **Which word in the passage tells you how the wind sounds?**

whistles

Writer's Companion ▪ UNIT 6
Lesson 3 *Vivid Verbs and Specific Nouns* 170

SHARING AND DISCUSSING

Divide students into small groups and give them word cards with nouns such as *wind, rain, snow,* and *sun.* Have each group work together to come up with vivid verbs that can follow the nouns and show clear and interesting action. For example, *wind howls, rain drizzles, snow drifts,* or *sun blazes.* As a class, discuss the noun/verb phrases of each group.

Explore Vivid Verbs and Specific Nouns

OBJECTIVES

- To deepen understanding of vivid verbs and specific nouns
- To identify and explore vivid verbs and specific nouns that clarify meaning

Teach/Model Use the graphic organizer at the top of the page to help students understand that vivid verbs and specific nouns paint a clear picture for readers. Provide an example of a vivid verb, such as *sparkled,* and a specific noun, such as *microwave.*

Guided Practice Read the directions for Part A and the example aloud. Discuss how the word *pet* is too general, while the word *dog* is clear and specific. Guide students in completing Part A.

Independent Writing Practice Read the directions for Part B aloud and have students complete the section independently. For Part C, see Writer's Strategy.

Name ______________________________

A Closer Look at Writer's Craft

Explore Vivid Verbs and Specific Nouns

Vivid verbs and **specific nouns** tell information in a clear and interesting way.

Vivid verbs paint clear pictures of action. — Using **vivid verbs and specific nouns** will tell your readers exactly what you want to say. — **Specific nouns** clearly tell your readers who, what, or where.

A. Read each sentence. Then rewrite the sentence by replacing the underlined noun with a specific noun. Answers will vary. Possible responses are given.

Example Jesse played with his furry <u>pet</u>.

Jesse played with his furry dog.

1. Belinda scooped up the <u>vegetable</u>.

   Belinda scooped up the carrots.

2. Could you please hand me the <u>dish</u>?

   Could you please hand me the bowl?

3. I could see the <u>feeling</u> in Grandpa's eyes.

   I could see the sadness in Grandpa's eyes.

B. Read these sentences from *Rocking and Rolling*. Rewrite the second sentence to make the underlined noun clearer.

Every year, there are 40,000 to 50,000 earthquakes that are strong enough to be felt. However, only about 40 of <u>them</u> are big enough to cause any damage.

However, only about 40 of the <u>earthquakes</u> are big enough to cause any damage.

C. Now imagine that you are writing to tell how one of the 40 earthquakes would feel and look. Write two new sentences that include vivid verbs and specific nouns. Use another sheet of paper. Answers will vary, but should offer two sentences that include vivid verbs and specific nouns about the earthquakes.

WRITER'S STRATEGY: EXPRESSING YOURSELF CLEARLY

Before students complete Part C, have them think about what an earthquake might feel and look like. Prompt students by showing magazine photographs, artwork, or a video of an earthquake.

Ask volunteers for examples of students vivid verbs and specific nouns. Create a word bank on the board that students can use as a resource for writing two new sentences.

Use Vivid Verbs and Specific Nouns

OBJECTIVES

- To prepare to write a play by choosing a topic
- To practice writing vivid verbs and specific nouns

Teach/Model Tell students that before they write a play, they should spend some time thinking about what emergency they will write about. Explain that by using vivid verbs and specific nouns, students will make their plays exciting, so they come alive. Read aloud the example as students listen. Discuss why the choices listed by the writer are effective.

Guided Practice Read aloud the directions for Part A. Then have students write verbs and nouns that could replace and clarify those that are given. Guide them in filling out the chart.

Independent Writing Practice Have students complete Part B independently. Invite volunteers to read some of the clear words they listed and explain why they chose them.

LESSON 3: VIVID VERBS AND SPECIFIC NOUNS

Name ______________________

Practice with Writer's Craft

Use Vivid Verbs and Specific Nouns

You are going to write a play about a problem. What will the problem be? What will your actors say? Write some verbs and nouns you plan to use. Then write new, clearer verbs and nouns that could replace unclear words. Here is how one student made lists of words to use.

Example

Unclear Words	Clear Words
Verb: falling	*tumbling, pounding*
Verb: open	*start, flip*
Noun: thing	*doorstep, branch*
Noun: clothes	*sweater, coat, hat*
Noun: place	*house, room, door, floor*

A. **Read the unclear words that could tell about an emergency such as an earthquake or flood. Then write clear verbs and nouns to replace them.** Answers will vary.

Unclear Words	Clear Words
Verb: looked	
Verb: came	
Noun: animal	
Noun: colors	
Noun: people	

B. **Choose some clear words from your chart. Use them to write the beginning of a play about an emergency such as an earthquake or flood. Do your writing on another sheet of paper.** Answers will vary, but should include some clear verbs and nouns from student's chart.

Writer's Companion ▪ UNIT 6
Lesson 3 *Vivid Verbs and Specific Nouns* 172

Reaching All Learners

BELOW LEVEL

Pair students who are writing below grade level with a partner. Tell the student to orally describe a natural disaster. Ask the partner to help the student write details that include vivid verbs and specific nouns. Have the partner explain how to write dialogue.

ADVANCED

Encourage students writing above grade level to write a play about an emergency with a heroic character based upon someone they know. Tell them to think about and include special qualities this character has and to show through dialogue these qualities.

ENGLISH LANGUAGE LEARNERS

Pair less fluent English speakers with more fluent English speakers. Have the more fluent English speakers supply appropriate words for vivid verbs and specific nouns their partner describes. Have the more fluent English speakers explain the words *dialogue* and *setting.*

The Parts of a Play

OBJECTIVES

- To understand how a play is developed and organized
- To analyze a Student Model

Teach/Model Read aloud the introduction. Explain to students that the writing model on this page is a first draft of a student's play. Read aloud the steps for how to develop a play. Explain that plays are written in a specific format. Then read the model aloud. As students listen, have them pay careful attention to the dialogue and the vivid verbs and specific nouns.

Guided Practice Point out the items below the Student Model. Read the first item aloud. Then guide students in finding the title of the play and drawing an arrow to it. Read the second item aloud and guide them in underlining the information that shows the play's setting.

Independent Writing Practice Have students complete Items 3–5 independently. Then discuss how vivid verbs and specific nouns emphasize the emergency in the play.

Use a Student Model

LESSON 3: VIVID VERBS AND SPECIFIC NOUNS

Name ______________________

Focus on the Writing Form

The Parts of a Play

A good **play** includes vivid verbs and specific nouns to introduce, develop, and solve a problem. Here is a third grader's draft of the beginning of a play. As you read, think about how the writer uses vivid verbs and specific nouns. Then answer the questions.

Student Model

DRAFT

High Water ←
by Clara
(A family kitchen as heavy rain falls outside.)

Draft the beginning. Be sure to introduce the setting and characters.

Oscar: Listen to the rain pound! Will it ever stop? Look outside. I can't even see the thing at the house across the street.

Mom: I think we should get ready to leave.

Grandmother: Let's (flip) on the TV first. The news can tell us how high the water is.

Introduce the problem. Describe what problem or situation your characters are facing.

Oscar: I tried. The TV won't work. I see water coming up to the top of the doorstep outside.

Mom: Let's call the people for help.

Grandmother: We can't. Our phone won't work.

Oscar: We need to go to the safe place.

Mom: Let's get the paper to figure out the best way to go.

Develop the play. Write about the events that happen. Use details to make the events interesting to read. Remember to use the proper format for writing a play.

1. Draw an arrow to the title of the play.
2. Underline the information that tells where the play is happening.
3. Draw two lines under the words that tell which character is speaking.
4. Circle the vivid verb the grandmother uses to ask Oscar to turn on the TV.
5. What is the problem in the play? The rain is causing a flood, and the characters need to get to a safe place.

Writer's Grammar
In a play, place a colon after the character's name before you write the words the character is saying.

WRITER'S GRAMMAR: THE CORRECT FORM OF A PLAY

Explain to students that a play has a certain form that is different from other kinds of literature, such as stories or poems. Tell them that before each character speaks in their play, students should place a colon after the character's name and before writing the words, or the *dialogue,* which the character says.

Evaluate a Play

OBJECTIVES

- To understand what to look for when evaluating a play
- To evaluate a play using a checklist
- To write an evaluation of a play

Teach/Model Discuss with students why it is important to evaluate writing. Explain that evaluating what they have written helps them see how to improve their writing. Discuss what to look for in evaluating a play, using the points at the top of the page.

Guided Practice Read the directions for Part A aloud and then read aloud the questions. Guide students in referring to the Student Model on page 173 for information. Tell students that using a checklist will help them focus on what they should evaluate. Point out the checklist in Part B and explain that it contains the writing skills practiced in this lesson.

Independent Writing Practice After reading aloud both sets of directions, have students complete Parts B and C independently. Point out the callouts on page 173 and explain that students can use them to help them evaluate the play.

LESSON 3: VIVID VERBS AND SPECIFIC NOUNS

Name ____________________

Evaluating the Student Model

Evaluate a Play

When you evaluate a play, ask yourself these questions:

- Does the writer use vivid verbs?
  (Look for strong verbs that give a clear picture of the action.)
- Does the writer use specific nouns?
  (Look for clear nouns that tell you exactly what the writer means.)

A. Reread the Student Model on page 173. Then answer these questions.

1. One of the sentences in the play includes a vivid verb that tells how the rain sounds. Write the sentence below.

   Listen to the rain pound!

2. Which sentence includes a specific noun that helps you picture how deep the water is? Write the sentence below.

   I see water coming up to the top of the doorstep outside.

B. Now evaluate the Student Model. Put a check in the box next to each thing the writer did well. Do not check the box if you do not think the writer did a good job.

- ☐ The play has a title.
- ☐ The play tells where the action is happening.
- ☐ The characters' words include many vivid verbs and specific nouns.
- ☐ The writer introduced and developed a problem.
- ☐ The writer used the correct form to write the play.

C. How do you think the writer could make the play better? Write your ideas below.

Possible response: The writer could replace the unclear verbs and nouns with vivid verbs and specific nouns.

Writer's Companion ▪ UNIT 6
Lesson 3 *Vivid Verbs and Specific Nouns* 174

See the rubric on page 207 for another way to evaluate the Student Model.

SHARING AND DISCUSSING

Explain to students that part of what captures readers' interest in a play is knowing that there is a problem that needs to be solved. Point out the problem in the play—a flood—and have students talk about why a flood can cause problems for people.

Revise by Adding More Vivid Verbs and Specific Nouns

OBJECTIVES

- To revise a play by adding more vivid verbs and specific nouns
- To revise writing by using clear words that strengthen its impact

Teach/Model Tell students that revising their writing gives them a chance to make changes that improve their work. Explain that in this activity, they will revise their writing by adding vivid verbs and specific nouns.

Guided Practice Read the introduction and the example aloud. Help students understand that the word *place* could be more specific. Discuss the images of comfort and protection that the word *shelter* conveys. Read aloud the directions for Part A. Guide students in completing Item 1 by finding other noun choices that are more specific, using the words in the Word Bank if they wish.

Independent Writing Practice Have students complete Items 2 and 3 of Part A on their own. Then read aloud the directions for Part B and have students complete it independently. Invite volunteers to share their revisions aloud.

LESSON 3: VIVID VERBS AND SPECIFIC NOUNS

Name ____________________

Revising the Student Model

Revise by Adding More Vivid Verbs and Specific Nouns

What is one thing the student writer could have done better? The writer could have added more vivid verbs and specific nouns. Here is an example of how a sentence from the Student Model can be improved.

Example We need to go to the safe place.

We need to go to the safe shelter.

Answers will vary. Possible responses are given

A. **Revise these parts of the Student Model. Replace each underlined word to make it clearer and more specific. Use words from the Word Bank if you wish.**

1. I can't even see the thing at the house across the street.
   I can't even see the fence at the house across the street.
2. Let's call the people for help.
   Let's call the police for help.
3. Let's get the paper to figure out the best way to go.
   Let's get the map to figure out the best way to go.

Word Bank
map
roof
newspaper
police
fence
neighbors
safe
shelter

B. **Revise the beginning of the play you wrote on page 172. Add vivid verbs and specific nouns. If you need more space, use another sheet of paper.**

Answers will vary, but should include vivid verbs and specific nouns. Play should be in correct form.

EXTENDING THE CONCEPT: UNDERSTANDING DRAMA

Explain that when writers create plays, they do so with the intention that their plays are going to be produced on stage (or in a movie). Tell students that the writers imagine the plays as they will be acted. Explain that in this way, plays are very different from other kinds of literature that are not meant to be acted out. Suggest that when they revise their work, students try to make their words exciting and lively, as though their plays will be acted on stage.

Review Word Choice

OBJECTIVES

- To review the terms *imagery, figurative language, vivid verbs,* and *specific nouns*
- To identify imagery, figurative language, vivid verbs, and specific nouns in literature

Teach/Model Read aloud and review the introduction. Tell students that their writing will convey pictures to the reader if they use carefully chosen words. Explain that in this lesson they will review what they have learned about choosing words. Read aloud the passage from *The Armadillo from Amarillo.* Have students listen for words that help the reader picture what is happening.

Guided Practice Ask students to think about the word choice in the poem. Help them to see how imagery, figurative language, vivid verbs, and specific nouns convey ideas that the reader can see. Guide them in completing Part B.

Independent Writing Practice Have students complete Part C independently by describing the color words that are used in the poem. Tell them to imagine that they are preparing to draw a picture based on the poem. Ask how these words would help them.

Use a Literature Model

LESSON 4: REVIEW WORD CHOICE

Name ______________________

Writer's Craft in Literature

Review Word Choice

Choosing words carefully can make your writing come alive. **Imagery** will help your readers "see" what you are writing about. **Figurative language** will paint a clear picture through comparing. **Vivid verbs** and **specific nouns** will tell your readers exactly what you mean.

A. Read the stanzas. Note how the writer has carefully chosen words.

Literature Model

From space the earth was a big round ball,
with swirling clouds of white
against a deep-blue background,
like the blue-black sky at night.

Planets shone around them,
reflecting starlike light.
In that silent room floating in the dark,
they traveled through the night.

—from *The Armadillo from Amarillo*
by Lynne Cherry

B. Read the stanzas again. Find examples of figurative language and other imagery.

1. Circle the figurative language that tells what the writer compares with the earth.
2. Underline the example of imagery that tells where the characters are as they watch the scene.

C. Which phrases from the poem include color words that paint a clear picture for the reader?

swirling clouds of white, deep-blue background, blue-black sky at night

Writer's Companion ▪ UNIT 6
Lesson 4 *Review Word Choice*
176

SHARING AND DISCUSSING

Before students complete Part C, explain that there are many ways to bring a poem's images to life. Tell them that color words can be made more interesting. For example, instead of using the word *red,* a writer might write the word *rosy* or *ruby-colored.* Give them other examples of interesting color words, such as *shiny* for *gold,* or *silver* or *sun-colored* for *yellow.*

Review Word Choice

OBJECTIVES

- To review the elements that make up word choice
- To review the concepts of imagery, figurative language, vivid verbs, and specific nouns
- To understand the effect of word choice on writing

Teach/Model Use the graphic organizer at the top of the page to remind students about different aspects of word choice. Give an example of imagery *(cottony snowball)*; figurative language *(a walking encyclopedia)*; a vivid verb *(spin)*; and a specific noun *(laptop)*. Remind students that good writers practice careful word choice.

Guided Practice Tell students that the example paragraph in Part A has carefully chosen words. Read the passage aloud. Then read aloud the directions for Part B and model finding figurative language in Item 1.

Independent Writing Practice Have students complete the rest of Part B and all of Part C independently.

LESSON 4: REVIEW WORD CHOICE

Name ______________________

A Closer Look at Writer's Craft

Review Word Choice

In this unit, you learned that careful word choices can make your writing clear and interesting.

Word Choice

- **Vivid verbs** show the reader clear action.
- **Imagery and figurative language** paint word pictures for readers.
- **Specific nouns** tell exactly who, what, or where.

A. Read this paragraph. Think about how the writer carefully chooses words.

Snowflakes fell to the ground as quietly as cat paws touching the grass. Reina gazed through her frosty window. She slipped into her coat and pulled on her boots. Then she dashed out into the icy wonderland.

B. Write two examples from the paragraph for each of the following. Possible responses are given.

1. Figurative language Snowflakes fell to the ground as quietly as cat paws touching the grass; icy wonderland
2. Vivid verbs fell, gazed, slipped, pulled, dashed
3. Specific nouns snowflakes, paws, grass, window, coat, boots

C. Write two sentences you might add to the paragraph. Choose your words carefully. Include figurative language, vivid verbs, and specific nouns.

Answers will vary, but should include figurative language, vivid verbs, and specific nouns.

WRITER'S STRATEGY: FINDING VIVID VERBS

Before students complete Part C, explain that they can look up synonyms for verbs in an electronic or printed thesaurus. Tell students that these are resources that will help them expand their vocabulary and their ability to create images for their readers. For example, other verbs that could be used in place of the word walk include: *stroll, march, stride, hike, move, step,* or *go.* Discuss with students which of these words seems to be the most vivid alternative to *walk.*

Review Word Choice

OBJECTIVES

- To prepare to write by selecting a topic and carefully choosing words
- To practice writing an invitation

Teach/Model Tell students that before they write an invitation, they need to think of some ideas and plan what they want to write. Read aloud the text at the top of the page and have students read the example.

Guided Practice Read aloud the directions for Part A. Then guide students in completing the chart with carefully chosen words.

Independent Writing Practice Have students complete Part B independently. Invite volunteers to read their opening sentences for an invitation aloud.

LESSON 4: REVIEW WORD CHOICE

Name ______________________

Practice with Writer's Craft

Review Word Choice

Before you write, you should think about making careful word choices. Here is how one third grader made careful word choices while preparing to write an invitation.

Example **Kind of Invitation:** *Surprise Party*

Nouns	Better, more specific nouns	Verbs	Better, more vivid verbs	Description	Better description, imagery
place thing person	park party Kia	comes say talk	appears shout cheer	no noise good food	as quiet as mice treats that will melt in your mouth

A. Use the chart to answer the following questions. Write your answers on the lines. Answers will vary.

1. What other vivid verbs could you use for *comes* and *say*?

____

2. What other imagery could you use for something that makes no noise?

____

____

3. What other imagery could you use for good food?

____

____

B. Write one to two opening sentences for an invitation to a surprise party. Answers will vary.

____

____

____

Writer's Companion ▪ UNIT 6
Lesson 4 *Review Word Choice*
178

Reaching All Learners

BELOW LEVEL

Help students who are writing below grade level by prompting them with questions, such as, "What event do you want to invite people to?" Then encourage them to follow the Student Model, but reword it with their own details.

ADVANCED

Encourage students writing above grade level to think about how to design an invitation that could have special graphics or three-dimensional features. Have them sketch a design that could be used for the invitation.

ENGLISH LANGUAGE LEARNERS

Pair less fluent English speakers with more fluent English speakers. Have students discuss their ideas for invitations and then help each other identify appropriate English words and phrases to include.

The Parts of an Invitation

OBJECTIVES
- To understand how an invitation is organized
- To analyze a Student Model

Teach/Model Explain to students that the writing model on this page is the first draft of a student's invitation. Ask them to read the steps for how to organize an invitation. Then read the model aloud. As students listen, have them pay attention to how the invitation is set up as a letter, with a heading, a greeting, and a closing.

Guided Practice Point out the items below the Student Model. Read the first item aloud and guide students in finding the heading of the invitation. Explain that the heading shows where the invitation comes from and the date on which it was written. Guide students in completing Item 2 by helping them find the greeting.

Independent Writing Practice Have students complete Items 3–4 independently. Then explain that carefully chosen words will have a positive effect on people who receive the invitation.

Use a Student Model

LESSON 4: REVIEW WORD CHOICE

Name ______________________________

Focus on the Writing Form

The Parts of an Invitation

A good **invitation** might include imagery, vivid verbs, and specific nouns. Here is an invitation written by a third grader. As you read, think about how the student chose words carefully and organized the invitation. Then answer the questions.

Student Model

DRAFT

85 Greens Road
Kerrville, Georgia 30340
October 23, 20--

Dear Ms. Cherry,

[Your book was like a bright light in our classroom.] It was so amusing. On March 3, we will have authors come to our school at 85 Greens road in Kerrville, Georgia. We would be happy if you could come.

Authors will speak to students. Then we will serve cookies and juice. We will sing and dance in a show to give our thanks.

It would be a dream come true if you could come to visit our school on March 3 for Author's Day. Thank you.

Sincerely,
Rosario

Write a heading and greeting. Write your address, the date, and a greeting for your invitation.

Draft the opening. Introduce the topic of your invitation.

Develop the topic. Include main ideas and details to describe the topic of your invitation.

Conclude. Remember to include details about when and where the event will take place. Close and sign your letter.

Possible responses are given.

1. Underline the heading of the invitation.
2. Circle the greeting.
3. Box two vivid verbs and two specific nouns.
4. Put [ ] around an example of figurative language.

Writer's Grammar
Remember to include the address in the heading of your invitation. What else is included in the heading?

179 Writer's Companion ■ UNIT 6
Lesson 4 *Review Word Choice*

WRITER'S GRAMMAR: THE FORM OF AN INVITATION

Explain to students that when they are writing an invitation, they should include their address in the heading of your invitation. Tell them they should also include the date on which the invitation is being written. Point out other important elements of this type of invitation, such as a polite greeting, a closing, and a signature. You may also want to add that invitations must include the date and time of the event, as well as a request to respond by a certain date.

Evaluate an Invitation

OBJECTIVES

- To understand what to look for when evaluating an invitation
- To see how a 4-point rubric is used to evaluate writing
- To find out why one Student Model received a score of 4 (excellent)

Teach/Model Tell students that some of the writing they do in school and on writing tests is evaluated using a special chart called a *rubric*. Explain that on the next two pages, they will learn how writing is scored using a rubric. Point out the rubric on page 181.

Guided Practice Read the directions for Part A aloud. Then read aloud the Student Model. Have students listen (without looking at the invitation) and think about why the Student Model is a success. Then have students read the model and the teacher's comments that go along with it. Explain that the comments show some of the things a teacher might look for when evaluating an invitation. Have students identify the comments that have to do with using a heading; carefully chosen words; using figurative language; focusing on the topic; and using a closing.

Use a Student Model

LESSON 4: REVIEW WORD CHOICE

Name ______________________________

Evaluating the Student Model

A. Two students were asked to write an invitation. This invitation received a score of 4. A score of 4 on a 4-point rubric means "excellent." Read the invitation and the teacher comments that go with it. Find out why this invitation is a success.

Student Model

DRAFT

62 Adams Street
Zane, Oregon 97402
February 10, 20—

Good job! You have included the correct heading.

Dear Officer Chin,

We are so glad that you watch out for the people in our town. Because you are so good at what you do, we would like for you to come and teach us more about safety. The Safety Fair will be at our school at 62 Adams Street on May 15. There will be people from all over the community coming to teach students about staying safe. We want to learn more about bike safety, walking home from school safely, keeping our homes safe, and many other things. Since we know that you are a walking encyclopedia when it comes to safety, we would love it if you could join us.

Yes! You have chosen words carefully to clearly tell about the event.

Excellent use of figurative language.

We will have many tables set up where students will visit to learn about safety. You can come to school early to have plenty of time to set up your table. We will have a lunch break at noon. There will be delicious sandwiches and the best fruit salad in town.

Nice work! You have developed the topic of the invitation.

We would be really happy if you could come to our school on May 15 for the Safety Fair. Thank you.

Sincerely,
Clark

Good work with giving a reminder in the closing.

Writer's Companion ▪ UNIT 6
Lesson 4 *Review Word Choice*
180

EXTENDING THE CONCEPT: MAKE IT SPECIAL

Tell students that an invitation offers an opportunity to let people know that they are important to you. Reread the invitation aloud. Have volunteers tell which parts of the invitation make the recipient feel that he is special and that it would be worthwhile for him to attend (such as *We are so glad that you watch out for the people in our town* and *Because you are so good at what you do)*. Discuss how students might make their invitations special.

Evaluate an Invitation

OBJECTIVES

- To find out why one Student Model that received a score of 2 (needs improvement)
- To use a rubric to evaluate an invitation

Teach/Model Tell students that they will read an invitation that received a score of 2 on a 4-point rubric. Read the Student Model aloud. Have students listen (without looking at the invitation) and think about why the invitation needs improvement.

Guided Practice Ask students to comment on any strengths and weaknesses they noticed as they listened to the invitation. Then have students read the Student Model and the teacher's comments. Guide them in identifying the areas in which the invitation needs improvement, such as the missing heading; the need for vivid verbs and specific nouns; and the missing conclusion and signature.

Independent Writing Practice Explain how to use the 4-point rubric at the bottom of the page. Ask students to complete it independently to show what score they would give the invitation.

Use a Student Model

LESSON 4: REVIEW WORD CHOICE

Name ______________________

Evaluating the Student Model

B. This paragraph got a score of 2. Why did it get a low score?

Student Model

DRAFT

Do not forget to include a heading with the address and date.

Dear Mr. Royer,

Good job with the greeting.

Would you come to meet us? We'd like it if you could show up early. Don't forget to bring all the things you'll need when you come.

Add vivid verbs and specific nouns to let Mr. Royer know about the event.

We'll have a good time, and there will be a lot of people here. Everyone will be happy to hear what you have to say.

Oops. Do not forget to conclude your invitation. Remind Mr. Royer about when and where he should come.

Sincerely,

Do not forget to sign your invitation.

C. What score would you give the student's story? Put a number on each line.

	4	3	2	1
Form ______	☐ The writer includes a clear heading, greeting, body, closing, and signature.	☐ The writer includes at least four of the following: clear greeting, body, closing, and signature.	☐ The writer includes three or fewer of the following: greeting, body, closing, and signature.	☐ The writer's form is not clear.
Development ______	☐ The topic and time and place of the event are clear.	☐ The topic is clear, but time or place is not clear.	☐ The topic is clear; time and place are not clear.	☐ The topic, time, and place are not clear.
Word Choice ______	☐ The invitation is filled with imagery, vivid verbs, and specific nouns.	☐ The invitation includes some vivid verbs and specific nouns.	☐ The invitation includes very few vivid verbs and specific nouns.	☐ The invitation includes no vivid verbs or specific nouns.

181

Writer's Companion ■ UNIT 6
Lesson 4 *Review Word Choice*

WRITER'S STRATEGY: IMPROVE WRITING THAT NEEDS HELP

Explain that the invitation on this page needs some changes. Invite students to improve the invitation by adding necessary details. In addition, suggest that they take the opportunity to make the event sound interesting. Tell students to use vivid verbs and specific nouns so the person who receives the invitation will want to come. After students have finished improving the invitation, ask them to explain the changes they made.

Extended Writing/Test Prep

OBJECTIVES

- To apply the craft of word choice in a longer piece of writing
- To select a topic for an extended piece of writing and begin the prewriting process
- To become familiar with the kinds of writing prompts that appear on writing assessments

Teach/Model Tell students that the next two pages will give them an opportunity to select a topic and follow the steps of the writing process as they complete it. Explain that they can choose from three possible writing activities. Review the choices listed in Part A, and have each student select one.

Guided Practice Explain that if students have another idea for an invitation that is not among the offered choices, they can write an invitation based on their own ideas. Then review the writing forms covered in theme. Have students reread the model tips in each lesson where the form was taught:

- A Poem: Lesson 1, Page 161
- A Thank-you Letter: Lesson 2, Page 167
- A Play: Lesson 3, Page 173
- An Invitation: Lesson 4, Page 179

Have students begin the prewriting process by completing Part B.

LESSON 4: REVIEW WORD CHOICE

Name ______________________

Extended Writing/Test Prep

Extended Writing/Test Prep

On the last two pages of this lesson, you will use what you have learned about word choice to write a longer written work.

A. Read the three choices below. Put a star by the writing activity you would like to do.

1. Respond to a Writing Prompt

   Writing Situation: Write an invitation for a surprise birthday party.

   Directions for Writing: Think about what information you will include in your invitation. Now write an invitation. Remember to include a heading, greeting, time and place, closing, and signature.

2. Choose one of the pieces of writing you started in this unit:

   - a poem (page 160)
   - a thank-you letter (page 166)
   - a play (page 172)

   Expand your beginning into a complete piece of writing. Use what you have learned about word choice.

3. Choose a topic you would like to write about. You may write a poem, a thank-you letter, or a play. Use imagery and figurative language, vivid verbs, and specific nouns.

B. Use the space below and on the next page to plan your writing.

TOPIC: ______________________

WRITING FORM: ______________________

HOW WILL I ORGANIZE MY WRITING: ______________________

____

____

____

Writer's Companion • UNIT 6
Lesson 4 *Review Word Choice* 182

SHARING AND DISCUSSING

Discuss invitations you have received to events (such as baby namings, weddings, birthday parties, family dinners, and so on). Explain that a carefully worded invitation shows that the writer cared enough about the recipient to spend time performing the writing process. Ask students whether they have received invitations that were particularly special, and if so, to explain why.

Extended Writing/Test Prep

OBJECTIVES

- To use a graphic organizer to plan writing
- To use the steps of the writing process to complete a longer piece of writing independently

LESSON 4: REVIEW WORD CHOICE

Name ______________________________

Extended Writing/Test Prep

C. In the space below, draw a graphic organizer that will help you plan your writing. Fill in the graphic organizer. Use the lines below to write notes and ideas for imagery, vivid verbs, and specific nouns.

Notes

____

____

____

____

____

____

____

____

D. Do your writing on another sheet of paper.

183

Writer's Companion ■ UNIT 6
Lesson 4 *Review Word Choice*

Teach/Model Explain to students that in Part C, they will select a graphic organizer to help them plan their writing. Provide each student with a copy of the appropriate graphic organizer (see back pages of this guide). Model how to complete it.

Guided Practice **Prewriting:** Help students as needed as they complete their graphic organizers. Remind them that a graphic organizer is a general guide to use as they write.

Independent Writing Practice **Draft, Revise, and Publish:** Have students complete their writing on a separate sheet of paper. After they have completed the writing process, ask them to think about how they might want to present their writing.

WRITER'S STRATEGY: CHOOSING THE RIGHT GRAPHIC ORGANIZER

Explain to students that different graphic organizers are used for different purposes. Tell them that they should use the graphic organizer that best fits the invitation, or other writing form they have chosen, and that suits their purposes. Share some examples of graphic organizers with them so they can visually understand the differences.

- A word web may help them brainstorm ideas for vivid verbs and specific nouns.
- A list might help them think of the parts of the writing form.
- A chart might help them organize their writing.

Answering Multiple-Choice Questions

OBJECTIVES

- To understand that some writing tests contain multiple-choice questions
- To develop the strategy of reading all answer choices before making a selection

Teach/Model Tell students that some writing tests include multiple-choice questions. Explain that this lesson will help them learn how to answer this type of question. Read aloud the directions for Part A. Review the test tip. Model reading all of the answer choices for Item 1 before choosing the sentence in which both contractions are correct. Then demonstrate how to mark the answer bubble.

Guided Practice Read aloud Item 2, and then have students read the three answer choices silently. Ask them to find the sentence with the correct form of all verbs. Have them mark the correct answer bubble under Item 2.

Independent Writing Practice Have students complete the last two items independently. If necessary, review the terms *contraction* and *apostrophe* before they complete item 3.

LESSON 5: WRITING TEST PRACTICE

Name ______________________________

Writing Test Practice

Answering Multiple-Choice Questions

Some writing tests have questions with answer choices. This lesson will help you practice this kind of test.

A. Some test questions may ask you about sentences. Read the test tip. Then answer each question. Fill in the circle next to the best answer.

Read and answer questions 1-4.

1. Which sentence is correct?

- Ⓐ I'ts on the table near the bench where they're sitting.
- Ⓑ (filled) It's on the table near the bench where they're sitting.
- Ⓒ Its' on the table near the bench where the'yre sitting.

2. Which sentence is correct?

- Ⓕ We taked our coats when it was time to go outside.
- Ⓖ (filled) We took our coats when it was time to go outside.
- Ⓗ We took our coats when it were time to go outside.

3. Which sentence is correct?

- Ⓐ (filled) You're the one she's looking for right now.
- Ⓑ Your the one shes' looking for right now.
- Ⓒ Youre' the one she's looking for right now.

4. Which sentence is correct?

- Ⓕ You was so happy when she helped you.
- Ⓖ You were so happy when she help you.
- Ⓗ (filled) You were so happy when she helped you.

Test Tip: Remember that the apostrophe in a contraction usually goes in the place of the missing letter or letters.
Example:
he + is = he's

Writer's Companion ■ UNIT 6
Lesson 5 *Writing Test Practice*
184

SHARING AND DISCUSSING

Review the incorrect answer choices for Items 1 and 3. Explain why each answer choice is incorrect. Create a list of phrases on the board, such as *I am; she is; we are;* and *you are,* and ask volunteers to form contractions for each phrase.

Answering Multiple-Choice Questions (cont.)

OBJECTIVES

- To understand that some multiple-choice questions involve first reading a passage
- To become familiar with reading a passage in which sentences are numbered
- To practice using a multiple-choice format that requires using an Answer Sheet

Teach/Model Tell students that some multiple-choice items ask them to read a passage and then answer questions. Read aloud the directions for Part B and have them read the passage silently. Explain that the sentences are numbered because the questions refer to the sentences by number. Read aloud the first item. Model looking back at the passage to see whether a change is needed. Model marking the answer on the Answer Sheet.

Guided Practice Guide students in rereading the sentences to complete Item 2. Suggest that they read the test tip to review how some verbs are different from others.

Independent Writing Practice Have students complete Item 3 independently.

LESSON 5: WRITING TEST PRACTICE

Name ______________________

Writing Test Practice

B. Some test questions may ask you to read a passage and answer questions about it. The sentences in the passage may be numbered. Read the test tip, introduction, and the passage that follows. Then answer the questions on your Answer Sheet.

Maria is in the third grade. She wrote this story for her Language Arts class. She wants you to help her revise and edit the story. Read Maria's story and think about the changes she should make. Then answer the questions that follow.

Play at the Park

(1) She go to meet her friends when they're at the park. (2) They are all so happy to find out sh'es on the way. (3) They can't wait to see her. (4) Everyone plays on the slides and swings. (5) They have such a good time, and they say they will come again tomorrow.

1. What change, if any, should be made in sentence 1?

A Change ***go*** to **goes**

B Change ***to*** to **too**

C Change ***they're*** to **the'yre**

D Make no change

2. What is the BEST way to revise sentence 2?

F They are all so happy to find out shes' on the way.

G They be all so happy to find out she on the way.

H They are all so happy to find out she's on the way.

J No revision is needed.

3. What change, if any, should be made in sentence 3?

A Change ***can't*** to **cant**

B Change ***wait*** to **waits**

C Change ***see*** to **sees**

D Make no change

Test Tip: Some verbs, such as *have*, *go*, and *take*, do not follow rules you have learned. You must learn the correct forms of these verbs by memory.

Answer all test questions on this Answer Sheet.

1. Ⓐ Ⓑ Ⓒ Ⓓ
2. Ⓕ Ⓖ Ⓗ Ⓙ
3. Ⓐ Ⓑ Ⓒ Ⓓ

185

Writer's Companion ■ UNIT 6
Lesson 5 *Writing Test Practice*

ASSESSING STUDENT RESPONSES

If students are consistently answering test items incorrectly, try to determine why they are having trouble. One possibility is that they are misreading the answer choices.

Ask the student to demonstrate the process by which he or she answered the question. Review the answer choices together. Guide him or her in understanding why one choice is correct and the others are incorrect.

Explain that each question refers to a specific sentence. Then assist the student in answering the question that refers to that sentence

Answering Multiple-Choice Questions (cont.)

OBJECTIVES

- To practice completing multiple-choice questions that involve reading a passage
- To practice identifying errors in a passage

Teach/Model Explain that all of the questions on this test ask students to decide the correct way to write sentences. Read aloud the directions for Part C and have students read the passage silently. Then model selecting the correct answer choice for Item 1 and marking the answer.

Guided Practice Guide students in answering Item 2. Show them how to carefully read aloud the sentence and think about how to correct it. Provide a sample sentence as practice. Guide them in selecting the correct answer choice.

Independent Writing Practice Have students complete Items 3 and 4 independently. For Item 4, suggest that they review the test tip.

LESSON 5: WRITING TEST PRACTICE

Name ______________________________

Writing Test Practice

C. Some test questions may ask you the best way to revise or correct sentences in a paragraph. Read the test tip. Then answer each question. Fill in the circle next to your answer.

The following is a rough draft of a student's paragraph. It may contain errors.

Feeling Better

(1) My sister have been home sick this week. (2) Her teacher asked me when she'd be comeing back to school. (3) I sayed that she was feeling better than she'd felt earlier in the week. (4) The teacher said she hoped my sister would get better quick. (5) I tooked my sister's books to her and put them in her room. (6) I think she'll be ready to do her homework soon.

Test Tip: Ask yourself if the action in a sentence is happening now or if it happened in the past. This will help you choose the correct verb form.

1. Which underlined word is not used correctly?
   - Ⓐ My sister have been
   - Ⓑ teacher asked me how
   - Ⓒ better than she'd felt earlier
   - Ⓓ think she'll be ready
2. Which is the correct way to write sentence 2?
   - Ⓐ Her teacher asked me when shed be coming back to school.
   - Ⓑ Her teacher asked me when sh'ed be coming back to school.
   - Ⓒ Her teacher asked me when she'd be coming back to school.
   - Ⓓ Her teacher asked me when she'd be comeing back to school.
3. How should the underlined word in sentence 4 be correctly written?
   - Ⓐ quickly.
   - Ⓑ quickliest.
   - Ⓒ quickest.
   - Ⓓ Leave as is.
4. Which is the correct way to write sentence 5?
   - Ⓐ I taked my sister's books to her and put them in her room.
   - Ⓑ I took my sister's books to her and puts them in her room.
   - Ⓒ I tooked my sister's books to her and put them in her room.
   - Ⓓ I took my sister's books to her and put them in her room.

Writer's Companion ▪ UNIT 6
Lesson 5 *Writing Test Practice*
186

Reaching All Learners

BELOW LEVEL

Help students restate what the question asks them to do. For example: "Question 1 is telling me that an underlined word is not used correctly. I have to look at each answer choice to decide which one has the incorrectly used word."

ADVANCED

Encourage students to write sentences that are incorrect in their use of verb forms and give them to a partner. Suggest that they ask partners to tell them how to fix the verb forms in the sentences.

ENGLISH LANGUAGE LEARNERS

Read the passage aloud to the students. Have them follow along as they listen. Guide them in reviewing the answer choice for each question. Encourage students to listen to each answer choice to hear which one sounds correct.

Answering Multiple-Choice Questions (cont.)

OBJECTIVES

- To practice completing multiple-choice questions that involve reading a passage
- To learn to answer multiple-choice questions that involve filling in a blank
- To develop the strategy of reading all the answer choices before making a selection

Teach/Model Read aloud the directions for Part D. Explain that sometimes tests will ask students to fill in the blank in a sentence with a correct word. Remind them that contractions are formed by dropping a letter (or two) and adding an apostrophe. Model answering Item 1.

Guided Practice Guide students in answering Item 2. Read the sentence that contains blank (2). Explain that the phrase *last summer* shows that the action took place in the past. Model choosing the correct answer and marking it.

Independent Writing Practice Have students complete Items 3 and 4 independently. Suggest that they review the test tip before they answer Item 3. Then discuss their responses.

LESSON 5: WRITING TEST PRACTICE

Name ______________________________

Writing Test Practice

D. Some test questions may ask you to fill in sentences. Read the test tip. Then answer each question. Fill in the circle next to your answer.

Read "Camping." Choose the words that correctly complete questions 1–4.

Camping

We are a camping family! (1) ____________ gone camping every summer since I was little. Last summer, we (2) ____________ to a new campground. We could not drive very (3) ____________ because it rained during the whole drive. We finally arrived at the campground. We waited in the car until the weather cleared. Then we (4) ____________ our best to put the tent in the mud. We looked up to see a rainbow form. We will never forget that trip!

1. Which answer should go in blank (1)?

Ⓐ Wev'e
Ⓑ We've (filled)
Ⓒ Weve

2. Which answer should go in blank (2)?

Ⓕ go
Ⓖ goes
Ⓗ went (filled)

3. Which answer should go in blank (3)?

Ⓐ fastly
Ⓑ fast (filled)
Ⓒ faster

4. Which answer should go in blank (4)?

Ⓕ did (filled)
Ⓖ done
Ⓗ does

Test Tip: Read carefully to see if a word describes a verb. If it does, it is an adverb. Be sure to choose the correct adverb form. In the sentence *The boy skated slowly*, the word *slowly* is an adverb.

187

Writer's Companion ▪ UNIT 6
Lesson 5 *Writing Test Practice*

USING ACADEMIC LANGUAGE

Explain to students that most tests contain academic language that includes words and ideas they may not come across in their regular reading. Before asking students to complete the test items above, review the concept of filling in a blank. Explain that in other tests, they may actually be asked to write a word in the blank, but in this test, they are being asked to select the letter of the correct answer choice. Remind them that tests also have different ways of asking for answers to be recorded (such as using an Answer Sheet or marking a circle).

Writer's Grammar Guide

Parts of Speech

The parts of speech are the different kinds of words you use in sentences.

Nouns are words that name people, places, or things.

Common nouns name any person, place, or thing.

My **friend** moved here from a big **city.**

Proper nouns name a special person, place, or thing.

Oscar moved here from **New York City.**

Verbs are action words that tell what people, places, and things do or are like.

A verb's **tense** lets you know what time it is telling about.

Verbs in the **present tense** tell about action that is happening now.

Oscar **pushes** his bike up the hill.

Verbs in the **past tense** tell about action that has already happened.

Oscar **pushed** his bike up the hill yesterday.

Verbs in the **future tense** tell about action that will happen later.

Oscar **will push** his bike up the hill tomorrow.

Pronouns are words that can take the place of nouns.

A **subject pronoun** can take the place of a noun that is the subject of a sentence.

Oscar cared for the garden. **He** cared for the garden.

An **object pronoun** can take the place of a noun that receives the action of a verb.

Oscar helped Mrs. Lin. Oscar helped **her** with the garden.

A **possessive pronoun** takes the place of a noun that shows ownership.

Mrs. Lin's garden is filled with flowers. **Her** garden is beautiful.

Adjectives are words that tell about, or describe, nouns.

Cool breezes blew across the **lovely** garden.

Articles are special adjectives. *A, an,* and *the* are articles. They come before nouns.

Lin saw **an** ant on **the** top of **a** flower.

Writer's Grammar Guide

Prepositions are words that help a noun or pronoun join other words in the sentence.

We took the flowers **from** the garden. We gave them **to** a friend.

Conjunctions are joining words such as *and, but,* and *or.* Conjunctions take the reader from one thought to another.

Oscar gave seeds to Lin, **but** she grew her garden on her own.

Writer's Grammar Guide

Sentences

A sentence is a group of words that tells a complete thought.

Complete sentences have a subject and a predicate. The subject is the naming part of the sentence. The predicate is the action part.

The sun shines. (SUBJECT PREDICATE) The leaves and flowers grow on the plant. (SUBJECT PREDICATE)

Fragments are groups of words that do not form complete sentences. Some fragments do not have subjects. Others do not have predicates.

FRAGMENT: Big drops of rain (Needs a predicate.)
FRAGMENT: falling from the sky (Needs a subject.)
COMPLETE: Big drops of rain are falling from the sky.

Run-on sentences are two or more sentences that run together into one sentence. They need to be divided into separate sentences.

RUN-ON: Flowers bloomed they were beautiful.
RUN-ON: Flowers bloomed, they were beautiful.
CORRECT: Flowers bloomed. They were beautiful.

Combining sentences can make writing more interesting and easier to read. Combine short, choppy sentences with a comma and words like *or, and, but.*

Oscar forgot to water the garden. Lin watered the plants for him.
Oscar forgot to water the garden, but Lin watered the plants for him.

Mrs. Lin grew flowers. Mrs. Lin grew vegetables.
Mrs. Lin grew flowers and vegetables.

Writer's Grammar Guide

Capitalization

Some words must be capitalized. A word that is capitalized begins with a capital letter.

Capitalize the first word in a sentence.

We drove to the lake.
The weather was sunny.
All of us were excited.

Capitalize proper nouns. These are words that name special people, places, or things.

I saw my dad, my grandma, and my friend hop into the car.
I saw **Dad, Grandma,** and **Ahmed** hop into the car.

We couldn't wait to head for the lake.
We couldn't wait to head for **Lake Gomez.**

Capitalize the street name in addresses.

We stopped at 123 Maple Street for lunch on the way.

Capitalize the pronoun *I*.

Ahmed and **I** sat in the back seat.
He knew **I** would want to read and play games on the way.

Capitalize the first word, the last word, and all important words in a book title.

I read part of ***Harry Potter and the Goblet of Fire*** during the drive.

Capitalize the title in a person's name.

I read about **Doctor** Ruiz in one of my books.

Writer's Grammar Guide

Punctuation

Punctuation marks are special marks used to make words and groups of words clear.

An end mark is used at the end of a sentence. It helps tell your reader what kind of sentence you are writing.

A **period** is used at the end mark of a **statement.** A statement tells something.

We arrived at the lake in the early afternoon.
Dark clouds filled the sky.

A **period** is used at the end of a **command.** A command tells someone to do something.

Keep an eye on the sky.
Watch for rain.

A **question mark** is used at end of a **question.** A question asks something.

How long would we have to wait to swim?
Would rain ruin our day?

An **exclamation mark** is used at the end of an **exclamation.** An exclamation shows strong feelings.

The moon came out!
Oh! It is beautiful!

Commas are used to separate words, parts of dates, or parts of sentences.

Commas are used after the **introductory words** in a sentence.

First, we jumped in the water.
Yes, I was the first one in the lake.

Commas are used to separate things in a list of three or more things.

We put potato salad, sandwiches, and juice on the picnic table.

Commas are used in a greeting and a closing in a letter.

Dear Ahmed,
Sincerely,
Pat

Commas are used to separate a month and date from a year.

We went to the lake on August 5, 2006.

Writer's Companion
Writer's Grammar Guide
192

Writer's Grammar Guide

Commas are used to separate the name of a city from the name of a state.

The lake is near Brownsville, Texas.

Apostrophes can be used to show ownership or to take the place of missing letters.

An **apostrophe** can show that someone has or owns something.

family's lunch	parent's keys	Pat's book
Grandma's car	Ahmed's game	bird's nest

An **apostrophe** can replace letters left out when a contraction is formed.

is not	are not	we would	they are
isn't	aren't	we'd	they're

We aren't going to eat too much. We'd like to swim again later.

Quotation marks are placed around the words someone says. The end mark usually goes inside the quotation marks.

Ahmed said, **"**I'd like to go back in the water now.**"**
I asked, **"**How long until we have to leave?**"**

Abbreviations are shorter ways of writing words. An abbreviation usually begins with a capital letter and ends with a period.

WORD:	Friday	August	Mister	Doctor	Street
ABBREVIATION:	**Fri.**	**Aug.**	**Mr.**	**Dr.**	**St.**

193
Writer's Companion
Writer's Grammar Guide

Writer's Grammar Guide

Usage

Usage tells people the correct way to speak and write.

Pural nouns are words that name more than one person, place, or thing.

Add -*s* to most nouns to form the plural.

SINGULAR:	horse	barn	farm	town
PLURAL:	horse**s**	barn**s**	farm**s**	town**s**

Ten **horses** live in two **barns** on our **farms.**

Add -*es* to nouns that end in *s, ch, sh,* or *x.*

SINGULAR:	bus	bench	brush	fox
PLURAL:	bus**es**	bench**es**	brush**es**	fox**es**

We will wait on bench**es** for the bus**es** to go visit the farm.

Change the *y* to *i* and add *–es* if a noun ends in *y*.

SINGULAR:	city	pony	puppy	penny
PLURAL:	cit**ies**	pon**ies**	pupp**ies**	penn**ies**

I saw ten pon**ies** and four pupp**ies.**

Possessive nouns show that someone has or owns something.

Add *'s* to a singular noun to form the possessive.

SINGULAR:	pony	friend	dog
PLURAL:	pony**'s**	friend**'s**	dog**'s**

The pony**'s** food was near the dog**'s** bone.

Add only an apostrophe (') to a plural noun that ends in *s*.

SINGULAR:	ponies	friends	dogs
PLURAL:	ponies**'**	friends**'**	dogs**'**

Some of my friends**'** books had pictures of ponies**'** barns.

Contractions are formed by putting two words together. Use an apostrophe to replace letters left out when a contraction is formed.

TWO WORDS:	is not	are not	we would	they are	will not
CONTRACTION:	isn't	aren't	we'd	they're	won't

Writer's Companion
Writer's Grammar Guide
194

Writer's Grammar Guide

Subject-Verb Agreement The verb in a sentence must agree with the subject.

A **singular subject** must have a **singular verb.** You usually add *-s* to the verb for the singular form.

The **girl likes** to visit the farm.

A **plural subject** must have a **plural verb.** Do not add *-s* to the verb for the plural form.

The **girls like** to visit the farm.

Adjectives and Adverbs can be used to compare people, places, things, and actions.

Compare with the correct form of an adjective.

Add *-er* to most adjectives to compare two people, places, or things.

The horse is tall**er** than the pig.

Add *-est* to most adjectives to compare three or more people, places, or things.

The horse is the tall**est** of all the farm animals.

Describe and compare with the correct adjective: ***good, better, best.***

Use ***good*** to describe one person, place, or thing.

Visiting the farm was a **good** field trip.

Use ***better*** to compare two people, places, or things.

Visiting the farm was a **better** field trip than visiting the zoo.

Use ***best*** to compare three or more people, places, or things.

Visiting the farm was the **best** field trip all year.

Compare with the correct form of an adverb.

Add *-er* to most adverbs to compare two actions.

The pony ran fast**er** than the dog.

Add *-est* to most adverbs to compare three or more actions.

The pony ran fast**est** of all.

Tell about most verbs with the adverb *well,* NOT with the adjective *good*.

The bus driver drove **well** on the field trip to the farm.

195
Writer's Companion
Writer's Grammar Guide

Writer's Grammar Guide

Double Negatives

Negatives are words such as *no*, *not*, *nothing*, and *never*.

Use only one of these words in a sentence.

WRONG: The hens **never** laid **no** eggs.
CORRECT: The hens **never** laid eggs.
ALSO CORRECT: The hens laid **no** eggs.

WRONG: **Don't never** forget to feed the chicks.
CORRECT: **Don't** forget to feed the chicks.
ALSO CORRECT: **Don't ever** forget to feed the chicks.

WRONG: He **doesn't** like to clean **no** barns.
CORRECT: He **doesn't** like to clean barns.
ALSO CORRECT: He **doesn't** like to clean **any** barns.

WRONG: The farmer did **not** do **nothing** last week.
CORRECT: The farmer did **nothing** last week.
ALSO CORRECT: The farmer did **not** do **anything** last week.

Irregular verbs have different forms than most verbs.

The verb *be*

PRESENT: ***am, is, are***

I am on the farm now. You **are** on the farm now. She **is** on the farm now.

PAST: ***was, were***

I **was** there earlier. You **were** there earlier. She **was** there earlier.

The verb *go*

PRESENT: ***go, goes***

I **go** to the barn today. He **goes** to the barn today.

PAST: ***went***

I **went** to the barn yesterday. He **went** to the barn yesterday.

The verb *do*

PRESENT: ***do, does***

I **do** chores today. She **does** chores today.

PAST: ***did***

I **did** chores yesterday. She **did** chores yesterday.

Writer's Grammar Guide

Tricky Words

There are some tricky words to learn about.

Homophones are words that sound the same but have different spellings and meanings.

to, too, two

Word	Meaning	Sentence
to	toward	I am going **to** the store.
too	more than enough; also	My dad is going, **too.** **Too** many people are shopping.
two	the number 2	We'll come back in **two** hours.

there, their, they're

Word	Meaning	Sentence
there	at that place	I can't wait to get **there.**
their	belonging to them	Many people are getting in **their** cars.
they're	they are	I'm glad **they're** leaving.

its, it's

Word	Meaning	Sentence
its	belonging to it	My dog needs a new tag for **its** collar.
it's	it is	**It's** time to pay for the collar and go home.

Watch for other homophones such as: *ant/aunt, ate/eight, eye/I, be/bee, blew/blue, cent/sent, hear/here, hi/high, hole/whole, know/no, meat/meet, pair/pear, read/red, right/write, road/rode, sea/see, son/sun, threw/through, weak/week, who's/whose, wood/would, your/you're.*

Writer's Grammar Guide

Homographs are words that are spelled the same but have different meanings.

can

Meaning	Sentence
able to	I **can** put the new tag on the collar.
container	Then I'll open a **can** of dog food.

bit

Meaning	Sentence
took a bite of	My dog **bit** into the food.
small piece	My dog only ate a little **bit.**

Spelling Tips

These tips will help you spell many of the words you use.

Short Vowel Sound

The short vowel sound is usually spelled with one vowel.

a: bat fast ***e:*** let when ***i:*** with little ***o:*** drop job ***u:*** cut funny

Long Vowel Sound

The long vowel sound is sometimes spelled **vowel** + **consonant** + **e.**

make shape wide smile joke tone tune flute

Soft *c* and hard *c*

The letter ***c*** can stand for the **hard *c* sound** or the **soft *c* sound.**

hard ***c:*** car cold cave soft ***c:*** cent face ice

The letter ***c*** is usually followed by *i, e,* or *y* if a word has the soft *c* sound.

place, rice, race, pencil, center, dance

Soft g and hard *g*

The letter ***g*** can stand for the **hard *g* sound** or the **soft *g* sound.**

hard ***g:*** go gate soft ***g:*** gym giant

Writer's Grammar Guide

The letter ***g*** is usually followed by *i, e,* or *y* if a word has the soft *g* sound.

cage, gentle, giraffe, huge, page, stage

Silent Letters

Silent *k* You need to remember which words begin with silent *k*.

knee knife knew knight knock

Silent *w* You need to remember which words begin with silent *w*.

wrap wreck write wrong wrote

Adding *–ed* or *–ing* to a word that ends in *e*

If a word ends in *e*, drop the *e* before you add *–ed* or *–ing*.

like liked nap napping bake baking

Spelling Words to Practice

These words can be hard to spell. Practice spelling them until you know them well.

about address again all right already always because busy could
enough friend once people receive sure together tonight trouble

Conventions

Proofreading Strategies

Our language follows **conventions,** or rules. We write in sentences. We end them with punctuation marks. We leave spaces between words, and we indent paragraphs. All of these conventions help readers understand what we write.

As you proofread your writing, you should check to make sure you have followed the conventions. These strategies will help you proofread:

Wait before proofreading.

Put your writing away for a while. Then come back to it. You may see new things.

Proofread in steps.

1. Look at your **sentences.** Are they complete? Are they written correctly? Are your **paragraphs** indented?
2. Check your **language use.** Do your subjects agree with your verbs? Have you used the correct forms of adjectives and adverbs? Have you followed the rules for **capitalization** and **punctuation?**
3. Last, check your **spelling.** Circle any words that look strange. Use a dictionary if necessary.

Proofread with a partner.

Two pairs of eyes are better than one. Your classmate may find mistakes you did not see.

Writer's Companion
Proofreading Strategies
200

Conventions

Proofreading Checklist

This checklist will help you as you proofread your work.

Sentences and Paragraphs

- ☑ Is every sentence complete?
- ☑ Does each sentence begin with a capital letter and end with the correct end mark?
- ☑ Is each paragraph indented?

Grammar and Usage

- ☑ Do your verbs agree with their subjects?
- ☑ Have you used the correct verb tenses?
- ☑ Have you used *I* and *me* correctly?
- ☑ Have you used the correct form of adjectives and adverbs that compare?

Capitalization and Punctuation

- ☑ Have you capitalized proper nouns and the pronoun *I*?
- ☑ Have you used commas, quotation marks, and apostrophes correctly?

Spelling

- ☑ Are you sure of the spelling of every word?
- ☑ Have you always used *there* and *their* correctly?
- ☑ Have you spelled plural nouns correctly?

Mark	Meaning
℘	delete text
^	insert text
↻	move text
¶	new paragraph
≡	capitalize
/	lowercase
◯	correct spelling

Technology

If you use a computer spell checker, remember that it cannot tell homophones apart. For example, the spell checker does not know whether you mean *here* or *hear.*

201
Writer's Companion
Proofreading Checklist

Presentation

Presenting Your Work

Sometimes you write for yourself. Most of the time, you write for other people. When you let other people read or hear your writing, you **publish** your work.

You can publish your writing in many ways. Here are some ideas.

Publishing Ideas for Any Type of Writing

- Read your writing aloud.
- Have a friend read it silently.
- Post it on a bulletin board.

Publishing Ideas for Descriptions and Poems

- Draw or paint a picture to go with your writing.
- Cut pictures from a magazine. Make a collage.
- Make up a dance to go with your writing.
- Find a piece of music to go with your writing. Make a tape recording in which you read while the music plays in the background.

Strategies Good Writers Use

- Use your best handwriting.
- Add drawings that help your readers understand and enjoy your writing.

Writer's Companion
Presenting Your Work
202

Presentation

Publishing Ideas for Stories

- Work with friends. Act out your story.
- Draw pictures to go with your story.
- Read your story aloud to another class.
- Make a class storybook.
- Send your story to a magazine.
- Mail your story to a relative far away.

Acting Out a Story

You can follow these steps to perform a story or a personal narrative.

Step 1
Plan how the people in your story should sound to the audience. What are their voices like? How do they say their words? Experiment with your voice.

Step 2
Find props for your story. You can use different kinds of clothing, pictures, and other items.

Step 3
Decide how you want to present your story. Do you want to read it just as it is written, or do you want act it out? You could even ask classmates to help you present your story as a play.

Technology

You can use the computer to make a neat copy of your work. Type the words the way you wrote them. Use ***Return*** and then ***Tab*** to start a new paragraph.

203
Writer's Companion
Presenting Your Work

Presentation

Publishing Ideas for Reports

- Add maps and pictures. Make a tabletop display.
- Make a poster for the classroom bulletin board.
- Teach your classmates about your topic.
- Put your report in the classroom library for others to read.

Publishing Ideas for Persuasive Writing

- Send a letter to the editor of your school paper.
- Give a speech to your class.
- Publish your ideas on your school's website.
- Read your work aloud. Take a poll. Find out who agrees with you.

Strategies Good Writers Use

- Give your work a title.
- Put your name on your work.
- Check your facts and proofread carefully before you send a letter in the mail or post work on a website.

Presentation

Giving an Oral Report

Strategies	Applying the Strategies
Make note cards.	• Write each main idea on a note card. Put your cards in order, and number them.
Practice.	• Give your talk to a friend or family member. Think about how to make your talk better.
Speak clearly and slowly.	• Speak more slowly than you do when you're just talking. Look at your audience. Remember that they can learn from you.

Strategies for Listeners

- Think about the speaker's main idea.
- Try to learn from what you hear.
- Ask questions to learn more.

Writer's Glossary of Terms

adjective: a word that describes a **noun**
adverb: a word that describes a **verb,** an **adjective,** or another **adverb**
compare and contrast essay: an essay that tells how things are alike and different
descriptive paragraph: a **paragraph** that tells what something or someone is like
detail: a fact, event, or statement; details usually tell about a main idea
dialogue: words spoken by characters in a story or play
directions: writing that tells how to do something
essay: a piece of writing that is not a story, usually with a clear purpose for writing
fact: something that is true
how-to essay: an **essay** that tells how to do something
letter: a written message to someone
letter of invitation: a letter asking someone to come to an event
main idea: what something is mostly about
notes: writing that records facts or ideas from a source of information
outline: a system for organizing information and notes
paragraph: a group of sentences with a single main idea or topic
paragraph of information: a paragraph that presents facts and data
paragraph of explanation: a paragraph that tells how something works or what something is like
paragraph that compares: a paragraph that tells how things are alike
paragraph that contrasts: a paragraph that tells how things are different
paragraph that explains: a paragraph that tells how something is done, what something is like, or how to do something
personal narrative: a story about the experiences of the writer
personal story: a story about what happened to the person who wrote the story
persuasive paragraph: a paragraph that tries to convince someone to do something or to think a certain way
play: a story written for the stage
poem: a piece of writing, often with rhyme
predicate: what the subject of a sentence does or is like
purpose for writing: the reason why someone writes something
reason: why something happens or is true
research report: a piece of writing that comes from study and from looking things up
review: a piece of writing that sums up and gives an opinion about a story, book, play, or movie
rubric: a guide for scoring or evaluating something
sensory detail: a detail that "speaks" to one of the senses—sight, sound, touch, smell, or taste
sequence: the order in which things happen
speech: a spoken message or statement
story: a made-up tale
subject: what a sentence is about
summary: a short piece of writing that wraps up the main points of a longer piece of writing
thank-you letter: a letter telling someone you liked and appreciated something they gave or did for you
topic: what something is about
verb: a word that names an action

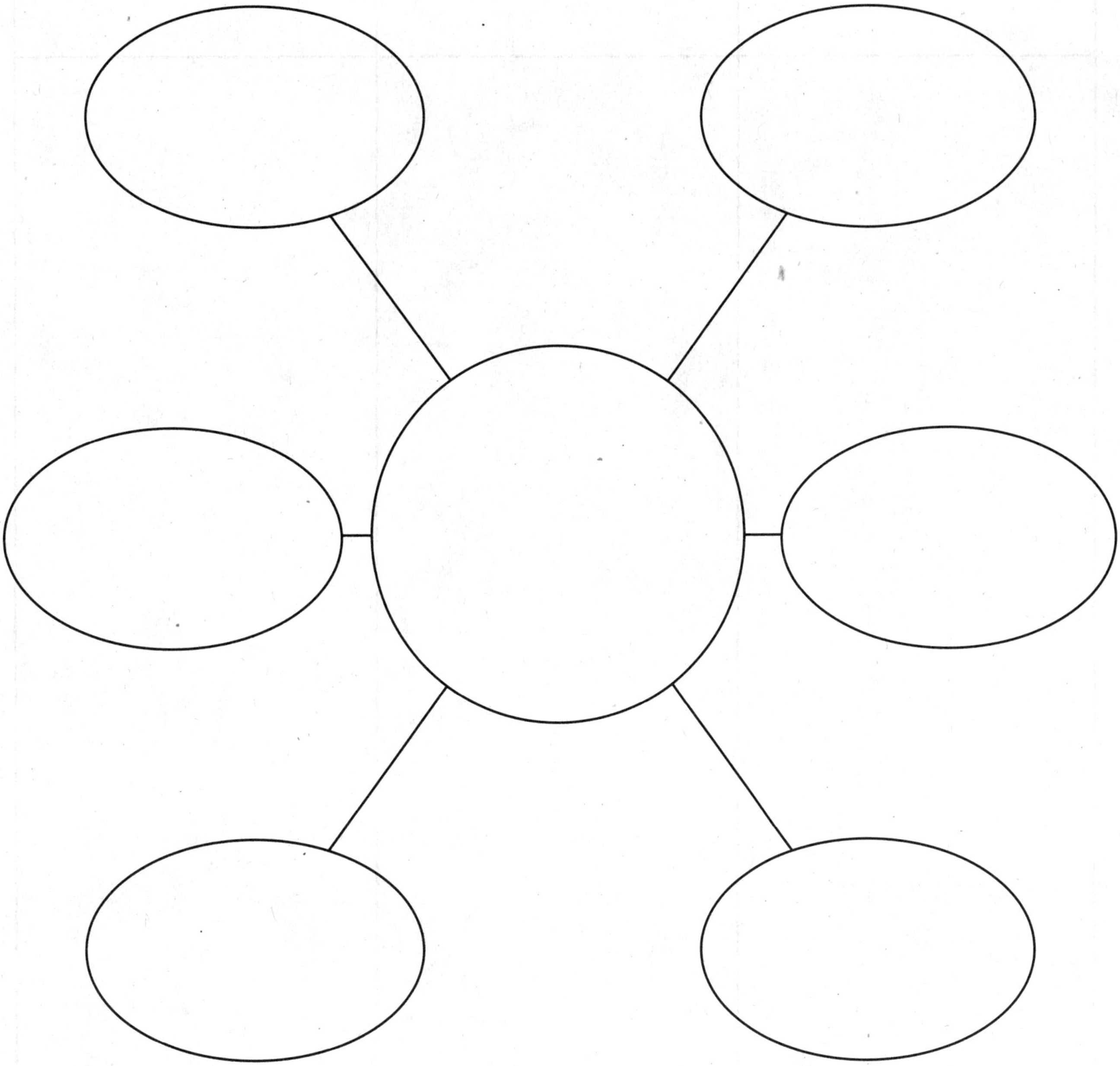

Three-Column Chart

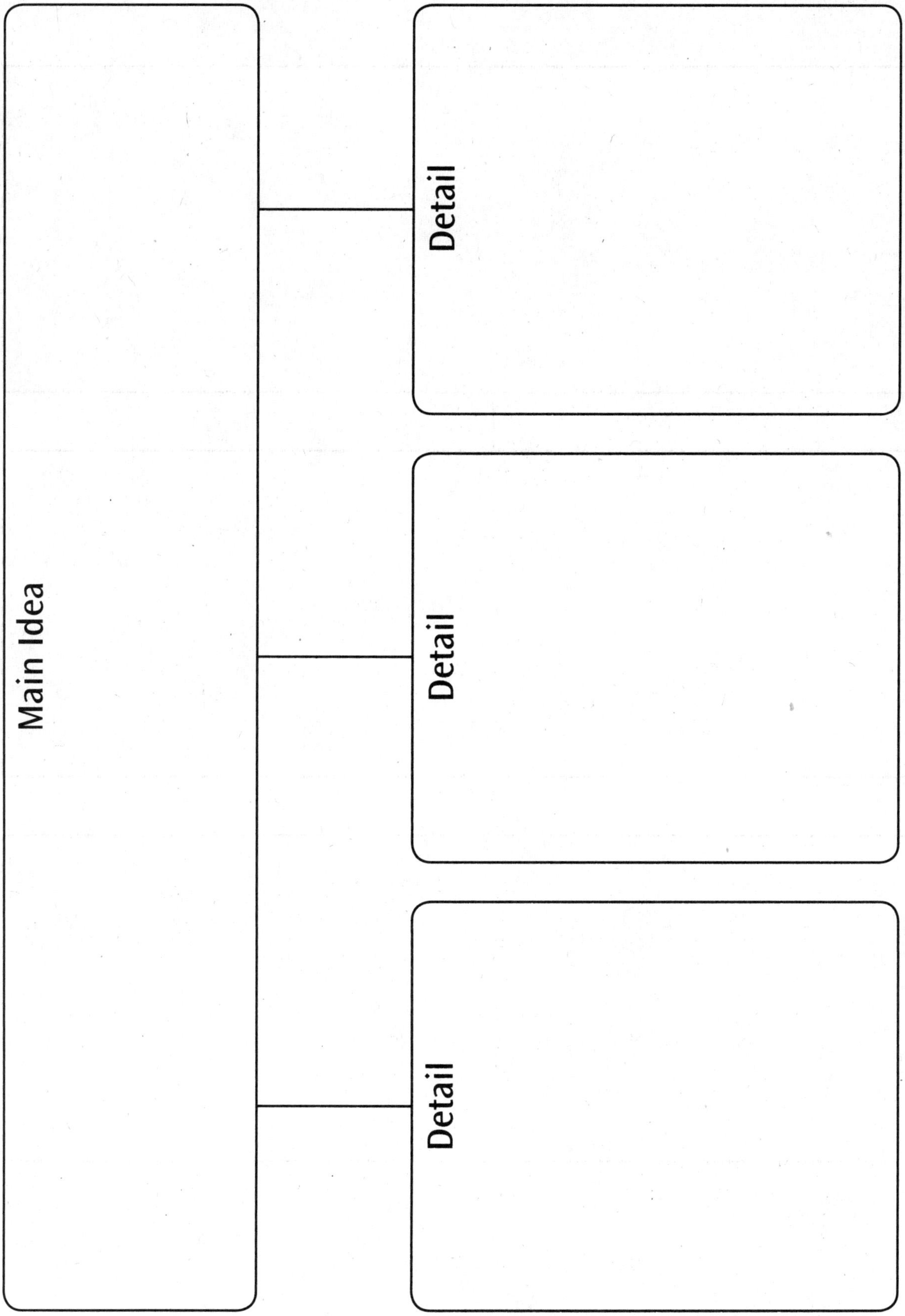
Main Idea
Detail
Detail
Detail

Flow Chart

Venn Diagram

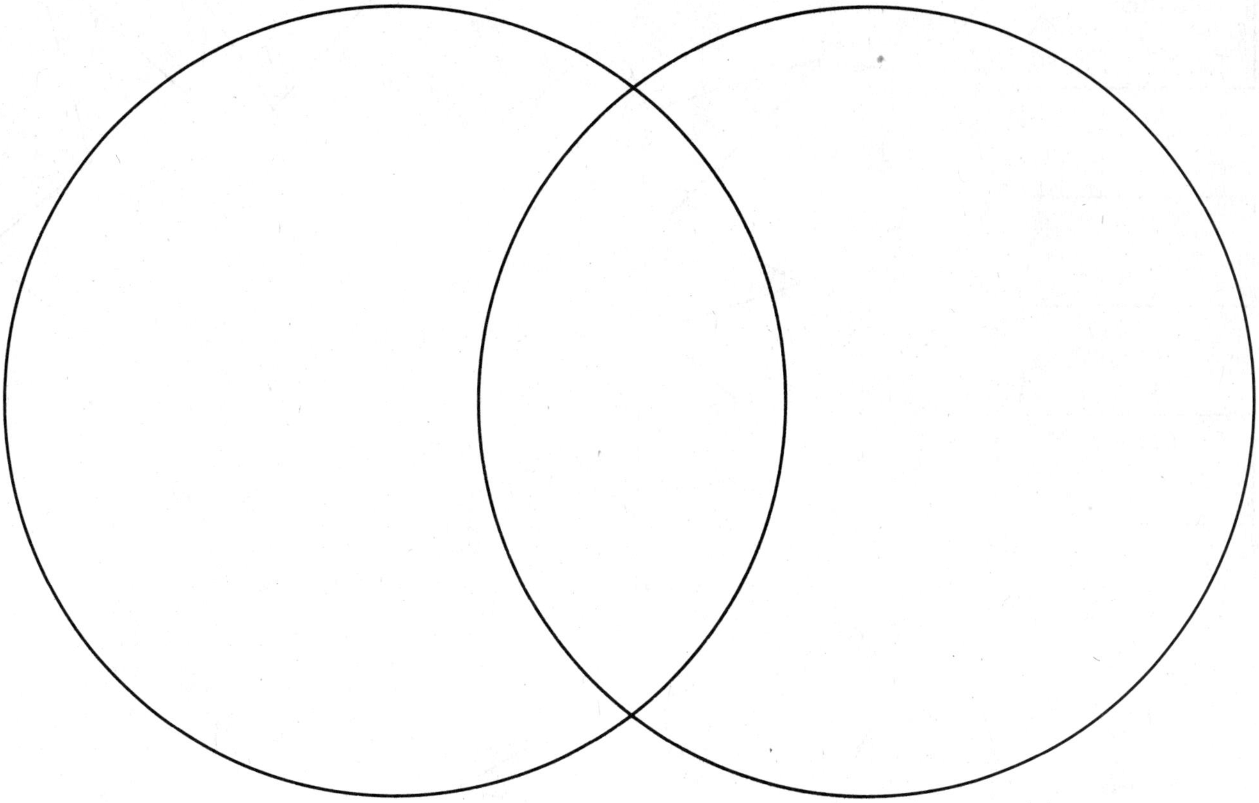

Cause-and-Effect Chart

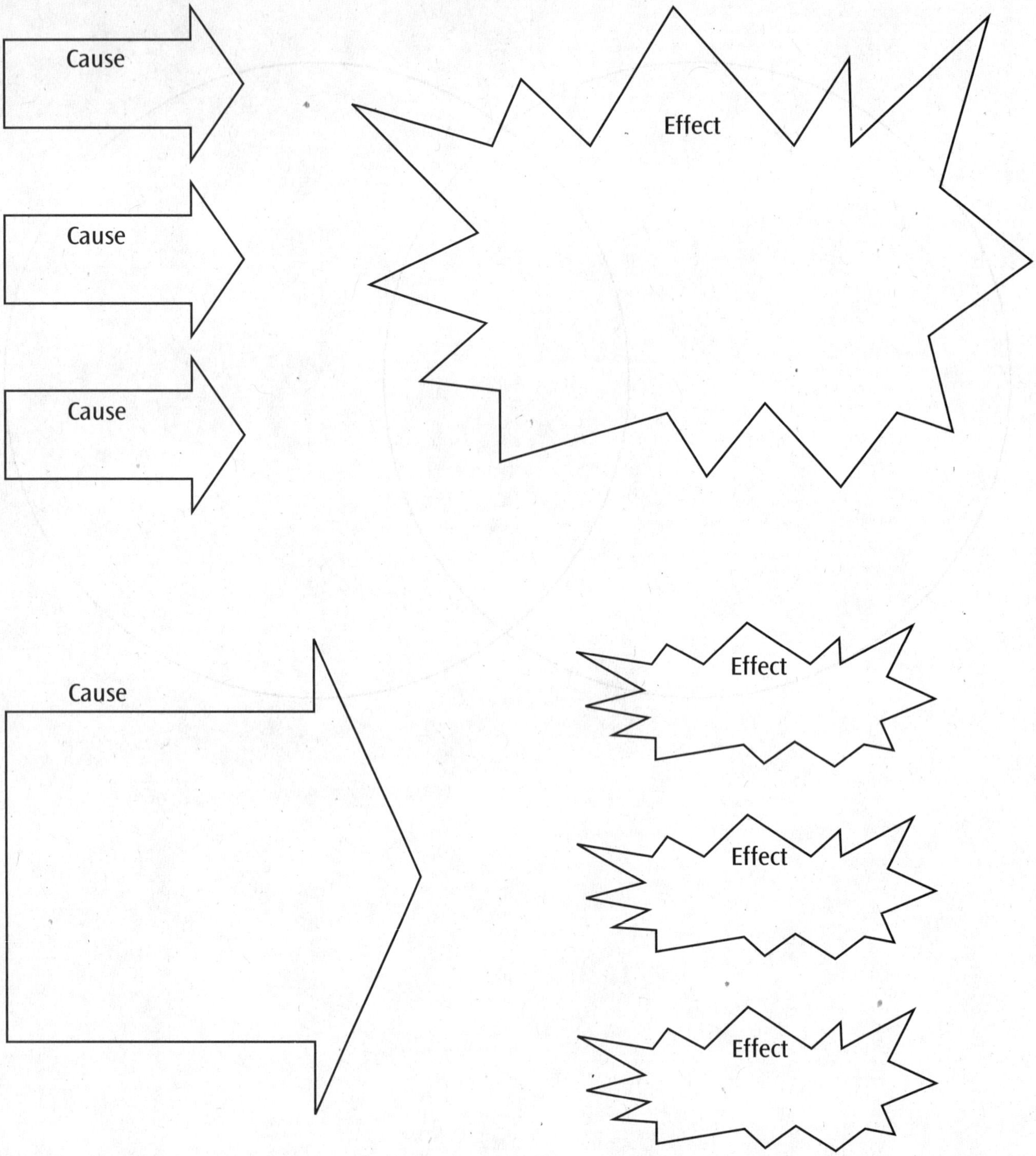

Using Student Rubrics

A rubric is a tool a teacher can use to score a student's work. A rubric lists the criteria for evaluating the work, and it describes different levels of success in meeting those criteria. Rubrics are useful assessment tools for teachers, but they can be just as useful for students. In fact, rubrics can be powerful teaching tools.

Before Writing

- When you introduce students to a **new kind of writing** through a Student Writing Model, discuss the criteria used on the rubric, and ask students to decide how well the Student Writing Model meets each criterion. Have students add to the rubric other criteria they think are important.
- Before students attempt a new kind of writing, have them focus on the **criteria** listed on the rubric so that they have specific goals to aim for.

During Writing and Editing

- As students are **prewriting** and **drafting,** have them refer to the rubric for important writing traits that they need to include in their writing.
- When students are ready to **revise,** have them check their writing against the rubric to determine if there are any writing traits that they can improve.

After Writing

- The rubric can be used by individuals to **score their own writing** or by peers or small groups for peer assessment. Students can highlight the parts of the rubric that they believe apply to a piece of writing.
- Students can keep the marked rubric in their portfolios with the piece of writing it refers to. The marked rubrics will help students **see their progress** through the school year. In conferences with students and family members, you can refer to the rubrics to point out both strengths and weaknesses.

Guidelines for Interpreting Holistic Scores

Holistic scoring has been used for many years to evaluate writing samples. Many state writing assessments, for example, employ some form of holistic scoring. In holistic scoring, the reader takes several features into consideration—weighing and balancing strengths in one feature against weaknesses in another—to arrive at a single, overall score.

A 4-point Scoring Rubric for Writing can be found in this section and in the back matter of the Student Edition. The rubric defines the characteristics of performance for each of the four possible scores. This rubric is similar to those used in many state writing assessments. Unfortunately, the numbers—1, 2, 3, and 4—carry no intrinsic meaning. To help interpret holistic scores to parents, other teachers, or administrators, use the following guidelines.

Score	Qualitative Label	Characteristics
4	Advanced	Students at this level exhibit superior performance. Their writing is thoughtful, organized, well developed, and fluent. They also exhibit consistent control of written language.
3	Proficient	Students at this level display solid performance. Their writing is mostly thoughtful and organized, moderately developed, and reasonably fluent. They may display a general control of written language.
2	Basic	Students at this level show uneven performance. Their writing is sometimes thoughtful and organized, mostly undeveloped, and lacking in fluency. They may display a lack of control of written language.
1	Limited	Students at this level exhibit very limited performance. Their writing is not thoughtful, often disorganized, usually undeveloped, and lacking in fluency. They may display great difficulty in controlling the conventions of written language.

What if your state or district uses a different rubric, say a 6-point rubric? Included in this section is a 6-point rubric for those teachers who may want to use a 6-point scale.

Regardless of which system your state or district uses, keep the following point in mind. The writing features being evaluated are more important than the numbers or scale used on the rubric (i.e., 4-point, 6-point). Ideally, the rubric used for classroom assessment should be aligned with the rubric used for the state assessment. For example, if the state rubric evaluates "development of ideas," then the classroom rubric should also evaluate this feature. The scoring rubrics used in the *Writer's Companion* have been designed to match the rubrics used in most state assessments.

Scoring Rubric for Writing

	Score of 4	Score of 3	Score of 2	Score of 1
	☆☆☆☆	☆☆☆	☆☆	☆
FOCUS/IDEAS	The paper is completely focused on the task and has a clear purpose.	The paper is generally focused on the task and the purpose.	The paper is somewhat focused on the task and purpose.	The paper does not have a clear focus or a purpose.
ORGANIZATION/PARAGRAPHS	The paper has a clear beginning, middle, and ending. The ideas and details are presented in logical order. The writer uses transitions such as *Finally, The next day,* or *However,* to show the relationships among ideas.	The ideas and details are mostly presented in logical order. The writer uses some transitions to show the relationships among ideas.	The organization is not clear in some places.	The paper has little or no organization.
DEVELOPMENT	The paper has a clear central idea that is supported by strong specific details.	The paper has a central idea and is supported by details.	The paper does not have a clear central idea and has few supporting details.	The central idea is not clear and there are few or no supporting details.
VOICE	The writer's viewpoint is clear. The writer uses creative and original phrases and expressions where appropriate.	The writer's viewpoint is somewhat clear. The writer uses some original phrases and expressions.	The writer's viewpoint is unclear.	The writer seems uninterested in what he or she is writing about.
WORD CHOICE	The writer uses clear, exact words and phrases. The writing is interesting to read.	The word choices are clear. The writer uses some interesting words and phrases.	The writer does not use words or phrases that make the writing clear to the reader.	The writer uses word choices that are unclear or inappropriate.
SENTENCES	The writer uses a variety of sentences. The writing flows smoothly.	The writer uses some variety in sentences.	The writer does not use much variety in his or her sentences.	There is little or no variety in sentences. Some of the sentences are unclear.
CONVENTIONS	There are few or no errors in grammar, punctuation, capitalization, and spelling.	There are a few errors in grammar, punctuation, capitalization, and spelling.	There are some errors in grammar, punctuation, capitalization, and spelling.	There are many errors in grammar, punctuation, capitalization, and spelling.

Scoring Rubric for Writing

	Score of 6	Score of 5	Score of 4	Score of 3	Score of 2	Score of 1
	☆☆☆☆☆☆	☆☆☆☆☆	☆☆☆☆	☆☆☆	☆☆	☆
FOCUS	The writing is completely focused on the topic and has a clear purpose.	The writing is focused on the topic and purpose.	The writing is generally focused on the topic and purpose.	The writing is somewhat focused on the topic and purpose.	The writing is related to the topic but does not have a clear focus.	The writing is not focused on the topic and purpose.
ORGANIZATION	The ideas in the paper are well-organized and presented in logical order. The paper seems complete to the reader.	The organization of the paper is mostly clear. The paper seems complete.	The organization is mostly clear, but the paper may seem unfinished.	The paper is somewhat organized, but seems unfinished.	There is little organization to the paper.	There is no organization to the paper.
SUPPORT	The writing has strong, specific details. The word choices are clear and fresh.	The writing has strong, specific details and clear word choices.	The writing has supporting details and some variety in word choice.	The writing has few supporting details. It needs more variety in word choice.	The writing uses few supporting details and very little variety in word choice.	There are few or no supporting details. The word choices are unclear.
CONVENTIONS	The writer uses a variety of sentences. There are few or no errors in grammar, spelling, punctuation, and capitalization.	The writer uses a variety of sentences. There are few errors in grammar, spelling, punctuation, and capitalization.	The writer uses some variety in sentences. There are a few errors in grammar, spelling, punctuation, and capitalization.	The writer uses simple sentences. There are some errors in grammar, spelling, punctuation, and capitalization.	The writer uses simple sentences. There are many errors in grammar, spelling, punctuation, and capitalization.	The writer uses unclear sentences. There are many errors in grammar, spelling, punctuation, and capitalization.